THE SPICE OF FLIGHT

FIRST EDITION
published in 2000
by
WOODFIELD PUBLISHING
Woodfield House, Babsham Lane, Bognor Regis
West Sussex PO21 5EL, England.

© Richard Pike, 2000

All rights reserved.
No part of this publication may be reproduced
or transmitted in any form or by any means,
electronic or mechanical, nor may it be stored
in any information storage and retrieval system,
without prior permission from the publisher.

The right of Richard Pike
to be identified as the author of this work
has been asserted by him in accordance with
the Copyright, Designs and Patents Act 1988

ISBN 1-873203-66-7

The Spice of Flight

RICHARD PIKE

Woodfield Publishing
BOGNOR REGIS · SUSSEX · ENGLAND

Acknowledgements

Friends have asked me recently if it has been my life-long ambition to write a book. The short answer, in plain truth, was 'no'. The notion for this book was born following a request from my wife's boss, The Bishop of Aberdeen and Orkney, The Right Reverend Bruce Cameron, to pen an article for his diocesan magazine. That item featured some dramatic flying in Kosovo, where I had worked as part of a team with the United Nations. Fortunately, the article was well received, and the idea of this book was conceived.

A famous author once described the writing of a book as a process which went through various stages. At first it was a hobby – a plaything. Then it became a mistress, then a master, and eventually it became a tyrant.

I would like to express thanks to my family for their support and positive encouragement, particularly when I reached the 'tyrant' stage.

This book is dedicated to my wife, Sue, and to our children Lizzie, Alan, and Sally.

Contents

Introduction 7
1   On Manoeuvres 9
2   Early Days 15
3   A Flying Scholar 25
4   Entering the RAF 30
5   Pilot Training 35
6   Gnats over Anglesey 41
7   Hunters at Chivenor 45
8   Lightning Conversion 50
9   Radar Sorties 57
10   Off to Pisa 63
11   Fun & Farewells 70
12   Welcome to Gütersloh 78
13   Missile Mayhem 87
14   Auf Wiedersehen Lightnings 93
15   Just Married 100
16   A Pilot Again 106
17   Phantoms & Flowers 111
18   A New Year Arrival 116
19   Chasing Bears 123
20   A Close Call 129

21	Phantoms in the Fog	134
22	Up with the Arrows	142
23	Tragedy Strikes	148
24	A Cloud on the Horizon	153
25	Farewell to Jets	157
26	Whirlwind Encounter	163
27	Called to the Rescue	168
28	In the Dunker	174
29	Learning from Lundy	180
30	Into the Mountains	187
31	Learning a New Language	194
32	To Germany Again	201
33	The White Tornado	208
34	Border Patrol	215
35	A Couple of Crises	223
36	Trouble in Tin City	233
37	Berlin – 1980	245
38	A Funny Old Life	253

Introduction

The characters depicted in this book have been based on real-life individuals. Many of the genre attracted to become members of Her Majesty's Forces have strong and colourful personalities. I hope this book's portrayal of some of them will be accepted with the tongue-in-cheek perspective intended.

During the book's composition, I have tried to take a generally light-hearted approach, and one which will be acceptable to aviators and non-aviators alike.

Some incidents, however, have not been treated in a jovial manner. Indeed, both the recollection and the documentation of certain events have caused the author considerable personal turmoil.

On some occasions within the narrative, specific episodes (not necessarily involving the author) have been used from time to time to illustrate and highlight particular points. These anecdotes have been based on fact, but the dramatisation effects have been left to the author's own resources.

If there is a serious facet of this book, it is to underline the skill and bravery of the men and women who devote their lives to the world of aviation.

CHAPTER 1

On Manoeuvres

"I feel ill," said Mike. He groaned slowly: "Ooohh!"

I looked at him. The appearance of his pale face and darkened eyes supported the statement.

"We all feel ill," I said.

"Ooohh." Mike groaned again. "Better get moving I suppose."

We lifted meagre kit onto our backs. The home-made rucksacks were roughly bundled together; the parachute material felt awkward to carry. Our boots, muddied and sodden, rubbed against blisters as we resumed our trudge through the forest. A light scattering of snow covered the forest floor.

Mike suddenly stopped. We both stood still, and listened intently. "What was that noise?"

We remained stationary as I beckoned Mike to crouch down. Mike held a finger to his mouth to indicate silence. The distant crunch of footsteps slowly became more audible. Fortunately it was nearly night; by chance our position gave us cover from the searching soldiers.

I pointed to a hut, barely visible in the gloomy half-light of the forest.

The soldiers headed for the hut, and we remained still for fear of giving away our position. We listened to the men as they tramped through the wooden hut, searching.

We heard the soldiers talk; snatches of German drifted to our hiding place.

"What are they saying?" Mike asked me in a whisper.

"*Niemand anders*…nobody else."

"Let's hope they sod off soon then," said Mike.

The soldiers searched around the outside of the hut as we continued to hide. Eventually, they decided to leave the abandoned hut. We waited until the retreating steps of the soldiers were out of earshot before we dared move again. A distant owl hooted; the noise added to the eerie atmosphere.

"Those guys'll be around for a while yet," I said. "Our best bet is to lie low here for an hour or two, and try to get some kip."

The survival exercise had been in progress for over a week already. We felt ravenous; our food for the last few days had comprised mainly hard-tack biscuits from survival packs. We had been shown how to trap and skin animals. Our Instructor, an ex-SAS man, had demonstrated the knack using a live rabbit. He had removed a kidney from the animal, still warm, then chewed and swallowed it in front of the class.

The Lech River, fast flowing and tempting, had caused some students sickness and diarrhoea in spite of the water purification tablets. Mike had attempted to clean his teeth in the river water, and had urged me to do the same. I had been doubtful.

With a drooping Mexican-style moustache, and normally with a ruddy complexion, Mike had a droll sense of humour; his reputation as a ladies' man tended to provoke his amusement. He had been a good companion; we had grown to know each other well during the extreme circumstances of our situation.

However, the tough conditions had got on our nerves; we had become edgy. The petty tooth-cleaning incident had caused tension between us.

The damp and discomfort had given us leg cramp. Our muscles ached. We felt weak and slightly dizzy. We had been in the same aircrew clothing, drab olive-green, throughout the exercise.

My survival partner and I both felt exhausted. We moved to an area deeper in the forest, beyond the hut, and found a place which offered scant shelter under some trees.

"This'll do," I said. Mike glumly surveyed the surrounding scene, and again complained of feeling ill.

We brought down the home-made kit bags from our backs, and wrapped ourselves in the flimsy parachute material. It was our bed for the night.

It was impossible to sleep properly; we both shivered violently with cold. Snatches of slumber were deceptive. Disturbed images, vivid and grotesque, filled the imagination.

A dream, frightening in its reality, swam through my subconscious mind: *A be-spectacled Officer had leered into my face.*

"You vill answer my questions, ya?"

I had remained silent. We had been lectured on the need for silence; it was our duty. The Officer had scowled at me. He brought his face closer to mine.

"Now. Pay attention. Vhy Ver you flying in German airspace?"

I continued to remain silent. The Officer looked at me intently. He was hostile and angry. He did not like my attitude; he did not like me.

Suddenly, I felt a surge of pain as he struck the side of my face with the back of his hand.

"Vhy make it difficult for yourself?" I felt the side of my face. It was swollen and sore. I thought it was cut, and looked for signs of blood.

The German Officer raised his voice: "I repeat my question. Vhat Ver you doing over German territory?"

I glared at the Officer. I said nothing. We were in a small room, sparsely furnished. The dull paint-work was peeling in places. An old clock was attached to one wall. A bare light-bulb hung from the ceiling.

"Come, come," the Officer had changed his demeanour. "I have no vish to hurt you. But you must answer my questions." We had been briefed on this approach. The Interrogator became 'Mr Nice Guy' all of a sudden.

"You tell me vot I vant to know," said the Officer, " and I can make life a lot better for you. Look at you. You're a mess! Just answer my questions, and I vill help you."

Still I remained silent. I looked at the old clock as its steady beat ticked routinely.

The Interrogator's attitude suddenly changed again. "Very vell," he said. "Have it your own vay. Guard… guard… guard…"

"Wake up! Stop dreaming." Mike was shaking me. "Seen the time? We'll miss the rendezvous." I hastily touched the side of my face. It had been scratched by the coarse surface of my forest pillow.

Quickly, we rolled up our parachute material, and slung the kit on to our backs as we set off past the trees. As we walked, we kicked up snow-covered leaves. In the icy dawn, the forest atmosphere felt frozen.

"Just a mile or so north of here there's a road," said Mike. "We might be able to hitch a lift. It's our only chance of making up the time."

We trudged through the thick forest, stopping occasionally to listen for *'Jager-corps'* soldiers. It was a matter of pride for them that they caught as many survival-exercise students as possible. After capture, the students were taken to an interrogation centre; my dream would have become reality. It was our aim, therefore, to avoid capture at all costs.

Eventually, we became aware of the road. Cautiously we checked that it was clear all around. Then we moved into a ditch alongside the road. Heavy frost covered the whole area. In the Autumn morning, the country-side felt still and peaceful.

"Not much traffic here," I said. We shivered in the cold, and scanned the road hopefully. A small group of children stood at the road-side; we were hidden from their view. We chewed hard-tack biscuits as we waited.

Suddenly, we heard the sound of a vehicle.

"It's a school bus," said Mike. "It's about to pick up those kids. You're the one with the German. Why don't you talk to the driver?" I looked at Mike's filthy clothing, and his mud spattered face, no doubt the mirror image of my own. He shrugged.

"Verzeihung!..." My spoken German was awkward and embarrassed as I apologised to the driver. The bus driver looked anxiously at our unshaven roughness, but he appeared to understand. Eventually, he jerked his head towards the back of the bus. Mike and I shuffled between the bus seats as we made our way to the rear. A few school children stared at us.

"Well. What do you know!" Mike said as we sat down. It was the first time I had seen him grin for days.

The bus continued on its country route, past various farm-houses, stopping occasionally to pick up passengers. Mike and I followed its progress on our map. As more impeccable schoolchildren boarded the bus, they gazed at the unusual passengers.

"Guten Tag!" we greeted each new-comer. Immaculately dressed, the children were formal and polite as they returned our greeting.

"What's that one saying?" asked Mike as he nodded towards an attractive young *frauline*.

"I think she's noticed that you stink of *shite*," I replied.

"Hmm," grunted Mike. "Charming. Not my fault, anyway."

"She doesn't know that."

Eventually, the bus was approaching a suitable drop-off point. Mike and I made our way to the front. The school-children continued to stare at us. I thanked the driver as we climbed down the bus steps.

*"Viel gluck…*good luck," said the driver.

"They're not such a bad lot really," remarked Mike. A few of the children waved back at us as the bus drove off.

We continued to make our way towards the rendezvous on foot, and aimed for tree-covered areas for protection from the prying eyes of the *'Jager-corps'*. The frost was melting in patches as the weak morning sun arose. We still felt cold, but at least the walking made us warmer.

"I reckon the rendezvous is in that wooded spot over there," I said eventually.

There was a stretch of open ground which had to be crossed. Keeping as low as possible, we ran across the open area. Soon we were within the protection of the woods. We advanced through the trees, and then saw a collection of Officers standing in the centre.

"You two've cut it a bit fine," said one of the Officers as he ticked our names on his board. "Anyway, well done. There's a hot drink over there."

Mike and I gratefully swallowed the hot tea.

"Well, well." said Mike after a while. "The things we do for Queen and Country." After a pause, he looked at me. "What made you join up anyway?"

His question took me by surprise. "Not for the money, that's for sure," Mike quipped.

For a moment, I took stock of our situation. I had not expected this.

It was Autumn 1979. We were both in the Royal Air Force. Mike was a helicopter crewman; I was a helicopter pilot. We had been sent to Bavaria in southern Germany for a survival exercise, and to learn techniques used in escape and evasion from enemy forces. We belonged to Number 18 Squadron, equipped with Wessex helicopters, and were based at Gütersloh in northern Germany.

My tired mind began to think back. I continued to sip the hot tea. Thoughts of the school-children on the bus tugged at my memory. I recollected my own school days.

"You were a fighter pilot once, weren't you?" Mike disturbed my pondering.

"Sure."

"You'll have to tell us about it some time."

"Maybe."

"It must have been quite exciting. A few hairy experiences, I'll bet."

"A few."

Strange ideas went through my senses. The exceptional circumstances had somehow stimulated my unusual line of thought. Annals from the past were suddenly swimming before me; I reflected on my flying career, and some of the intriguing and bizarre experiences. The meditations even provoked memories of my parents and grandparents.

Mike slumped down by a tree. He sat on the hard ground, and placed his mug of tea on a convenient rock. He looked up at me; I was still standing.

"I feel ill," said Mike. He groaned slowly.

CHAPTER 2

Early Days

Perhaps it had been his tough start in life which had fuelled my father's ambition.

"My parents had Victorian attitudes," said my father. His parents, married in 1898, had three children, all of them boys. The family had faced calamity when my grandmother had been widowed in her mid-30s: my paternal grandfather had fallen, cut his thumb badly, and consequently died of lock-jaw.

"My mother had a real struggle to bring us up on her own," said my father. His eldest brother, Roy, eventually became a scientist. He went to the Mount Wilson planetarium, in California, on an exchange programme. Within a few weeks of his arrival, he caught pneumonia and died. "Such a tragic loss of a brilliant mind," said my father. His other brother, Bill, joined the Army, and climbed the military hierarchy to become a Lieutenant General. "I was the youngest, and perceived to be a bit dim!" said my father ruefully.

Certainly, he was a non-academic, but my father still had a good brain, and the ability to have become highly focused. He was a fairly tall, good-looking man, with dark hair and penetrating eyes. He would have concentrated on a problem, and it would have worked and worried away in his mind. Eventually, when he had grasped the solution, he had the ability to explain it in a clear and simple way, no matter how complex the subject.

The up-bringing and education of my mother had been somewhat eccentric. "My mother and father would squabble and argue loudly," she said. "I suppose it was inevitable that they split up eventually." Her parents' separation was a dramatic blow for my mother and her two sisters. The education my mother received was largely thanks to her grandfather who was the headmaster of a private school. A graceful young woman with an artistic nature, my mother became a dancing teacher when her school days had ended. "In those times though, it was considered bad form for a lady to have a career," she said. "A woman's place was in the home." So when a dashing young fighter pilot appeared on the scene, who could have blamed her for accepting the proposal of marriage, even though she was still a teenager?

The black and white photograph of my parents' wedding day has been on the mantelshelf for years now. A handsome, shy couple could be seen staring out of the picture, as they held on to each other nervously. Their wedding was held on St George's Day, 1930.

Shortly after their marriage my parents were posted. My father became a Flying Instructor with the Central Flying School, based at Wittering, though the family lived at Henlow. My sister Caryl was born at Henlow on the 13th of April 1932, eleven years before me.

During that period, my father took part in some of the Central Flying School aerobatic displays which became well-known in the 1930s – fore-runners to the 'Red Arrows'. My parents were not there

CFS Aerobatic team 1930, RAF Wittering.

My parents' Wedding Day 1930.

very long, before they were sent off to Egypt. "We had to travel by ship, and it was such a traumatic journey, you wouldn't believe how sea-sick I felt," said my mother. "What made it worse was having to cope with Caryl, who was just a tiny tot," she sighed. "When we finally got to Egypt, we lived at Abu Sueir near Cairo. It was a really exciting and stimulating life."

My parents were posted back to England after their tour in Egypt, and just before the outbreak of World War 2, my other sister, Ann, was born.

For the early part of World War 2, my father was a night fighter pilot, and was credited with a number of 'kills' during war operations. During the Battle of Britain in 1940, he was in the Air Ministry. "From time to time, I tried to get home," he said, "but it was pretty hair-raising trying to travel across London. Usually, I just slept in my office." If he made a successful journey, it was to Wyntersbrook, a rambling old house which was my maternal Grandfather's home in Essex. "The damage to London was simply awful," said my father. "A smell of gas and burning was permanently hanging in the air." The efforts of rescuers, and the spirit of ordinary folk, always amazed and impressed my father.

After the Battle of Britain, my father had been the Station Commander at North Weald aerodrome, in Essex. The senior Army Officer there, Colonel Archie Crabbe, had asked to see him one time. "I must point out," the Colonel said, "that your over-exuberant and noisy fighter pilots are keeping my soldiers awake at night. It's interfering with their efficiency!"

There were a number of pilots at North Weald from the Air Forces of the Free World, for example from Norway, Poland, Canada, and Czechoslovakia. They were a wild lot. The discipline at North Weald had become a major problem. It showed the esteem the authorities held for my father that he was selected to sort it out. It was a measure of his ability that he succeeded.

Furthermore, my father and the Colonel remained firm friends for the rest of their lives.

"World War 2 undoubtedly helped your father's career," my mother had said. His success continued after the War. In the late

A portrait of my father as Chief of Air Staff, 1961.

1950s, he became the Commander-in-Chief of Fighter Command, and we lived in a fine house called 'Montrose', in Stanmore, Middlesex. The house had several staff, a gardener (useful in spite of his bad temper), a driver, and other trappings. In the garden we had a tennis court, and friends came round regularly. There was a happy atmosphere, despite the in-built formality of the situation. "Even at that stage in life," said my mother, "you were keen to become a pilot." Unfortunately, the attitude of my Headmaster (and even my parents) was not encouraging.

My education was in schools on the south coast of England. After preparatory school at Seaford, in Sussex, I moved a few miles down the road to Eastbourne College, "a cultural desert" as one of my cousins harshly described it. "You were tall, skinny, and spotty, with short-cut fair hair," said my sisters Caryl and Ann. They went on to describe a reticent nature which was fostered by a sheltered life-style. Nevertheless, I managed to make a few good friends at school, although I was evidently not over-impressed with some of my peer-group.

"You are like me," said my father. "You find academic studies a pain in the neck." At breakfast one morning, at Montrose, I was sitting with my father at the shiny dining-room table, knowing that the tricky subject of my school report was about to be broached. My father was not in a good mood. He had breakfasted, as usual, on porridge with syrup. He had the trick of re-uniting the syrup pot with its lid by inverting the syrup pot, and placing it slickly onto the lid which was lying on the table. He normally ensured that the domestic staff were not around before attempting his party trick. Occasionally, things went wrong. "Bugger...," he said under his breath that morning. A pool of syrup smudged the shiny table top.

My father got a pencil, and underlined certain words in the school report. He then slid the report across the table to me. On nearly every page he had underlined the words... *"fails to concentrate..."*

As a trained engineer my father decided to take on an unusual project. He built me a small car, with two gears, and powered by a two-stroke engine. I drove the car around the garden at Montrose, much to the gardener's anger and disgust. "You grew familiar with clutch control, gear-changing, and braking, and at an early age developed a sympathy with things mechanical," said my father. My hopes and dreams to have become a pilot were looking rather bleak on the academic front, but the small car helped to nurture a natural aptitude for driving.

After his job at Fighter Command, my father was promoted to the top position in the Royal Air Force, the Chief of the Air Staff. We moved from Montrose, to a flat in Hyde Park Gate, London, living just below the Sieff family of Marks and Spencer. Once or twice, we bumped into the Sieff's by the lifts. Other than that, there was no contact at all.

My sister, Ann, always the rebel, managed to smuggle a large Afghan Hound called 'Effie' into our London flat. After a few days, my father became suspicious, and summoned a member of his domestic staff. He asked for an explanation of certain happenings. The man became defensive, and eventually, incredibly, stormed off with the words: "I'm saying nuffing. Nuffing at all!" This was highly irregular; Chief's of the Air Staff were not used to such treatment.

'Effie's' skin, not to mention that of my sister and the member of domestic staff, was saved by an odd twist. The senior politician at the Air Ministry, a man called Hugh Fraser, husband of the author of historical novels, called at our flat. It soon became clear that Mr Fraser was more interested in Afghan Hounds than he was in matters to do with the Air Ministry. He spent ages playing with the dog, and evidently forgot the original reason for his call. After that visit, he regularly asked my father about 'Effie's' well-being.

Soon after the 'Effie' episode, my father invited me to see where he worked in the Air Ministry. I had to wait, like everyone else no doubt, in an outer room manned by half a dozen staff. Eventually, I was summoned.

"Sorry about that," said my father. "I had to deal with an Air Vice-Marshal who was upset at not being promoted."

"What did you say to him?"

"I shook him by the hand, and congratulated him on a brilliant career."

Later he let slip: "A good man would not have come to me like that. He should have understood that not everyone can be promoted right to the top."

My father brought us down-to-earth by opening the bottom drawer of his vast desk. He produced a tin of shortbread biscuits. "Only a lucky few are offered one of these," he said.

Usually, my father travelled in a special lift in the Air Ministry. "This time," he said, "we'll use the normal lift so we can stop halfway, and see another part of the building." We looked out over the River Thames, grey and murky, to the Shell Oil Company building opposite. Returning to the lift, we were joined by another man, who recognised my father immediately. I was amazed at how our lift companion suddenly seemed to waken up, and how he introduced himself as 'Group Captain so-and-so', and they had last met at 'such-and-such'. My father, polite and urbane, acknowledged everything, but later admitted to me that he had not the faintest idea who our lift companion was.

At about the time of this visit, my parents attended a dinner in London at which Her Majesty Queen Elizabeth The Queen Mother was the Guest of Honour. My father, sitting next to the Royal Guest, was aware that The Queen Mother was a music lover. However, my father was regretfully a musical ignoramus. Nevertheless he brought the subject into conversation that evening.

"It was so embarrassing," my father said to me later, after the dinner.

"Why?"

"Well, I mentioned that I liked Handel's music."

"Oops!"

"The Queen Mother immediately enthused. She was on about Zadok the Priest and other works. Then she asked me why I liked Handel."

"And?"

"I replied," said my father bashfully, "that I found him easy to grasp."

"Nice one!" I said. "Did The Queen Mother understand the joke?"

"She was *not* amused," said my father.

Some years later, when my father was President of the Royal Air Forces Association, he again sat next to The Queen Mother at a dinner in London.

"Did you mention anything about that dude Handle losing his grip?" I asked my father on his return home.

"Definitely not," he said.

During his era as Chief of the Air Staff, my father told me one time about a problem involving the V-bombers. He never told me anything classified 'secret', of course. He was highly discreet, and we learnt later in our lives that he had been worried that our flat at Hyde Park Gate was 'bugged'. However, he mentioned in general terms that Avro Aviation, the makers of the V-bombers, needed the dimensions of a nuclear bomb manufactured in the USA. Then the bomb-bay

My Father with the Queen Mother and Douglas Bader.

could have been built to exact limits. "The Americans," said my father, "are refusing to supply the information. They are worried about 'State Security'."

Fortunately, my father got on well with General Curtis LeMay, his opposite number in the United States Air Force. The two men were of similar temperaments, with focused minds, poor at small talk, a dry sense of humour, and very direct. They both knew struggles in their early lives, and possibly as a result, they were renowned for being kind to the 'small man'. A certain charisma marked them as leaders. Sometimes it was one extreme or the other: they were either liked or loathed. They were feared by many with whom they worked, and both men were 'hawks'. It was rumoured that General LeMay, some years later, had tried to persuade President Johnson to drop a nuclear bomb during the Vietnam War.

In order to resolve the bomb-bay impasse, a trip to the United States was organised for my father.

"Tom," said the US General to my father, "I'll need you to stand over here please."

My father obliged.

"Now, hold out your arms," said General LeMay. "Bend your knees a bit. Curl up the fingers of one hand."

"Shall we dance?" asked my bemused father.

"Never mind that," said the General. "You now have the dimensions you require!"

On that tenuous basis of measurement, the skilled draughtsmen and machine operators at Avro Aviation carried out their work to the n-thousandth of a millimetre.

In spite of the fascination of life when my father was Chief of the Air Staff, my parents both agreed that probably the most singular defining factor in their lives had been World War 2. In the middle of the War, a personal aspect had affected their lives. My mother became pregnant. My parents decided to abandon Wyntersbrook, and to set-up home in a house called Benbow Cottage, in the heart of war-torn Middlesex. I was born in the summer of 1943. Hitler was ranting, my father was fuming, and I was stuck in Benbow Cottage.

CHAPTER 3

A Flying Scholar

Sixteen years later, Hitler had committed *hara-kiri*, my father had calmed down a bit, and I was travelling on the A11 road to Cambridge. I was driving a fairly decrepit green motorbike, with 'L' plates front and back. The year was 1959. Rock and Roll abounded; Elvis was 'King'. My father was about to become the Chief of the Air Staff. Fidel Castro was leading the revolution in Cuba.

In that year the Royal Air Force awarded me a 'Flying Scholarship', a scheme which gave youngsters the opportunity to learn to fly. My chosen Flying Club was Marshall's of Cambridge. The Club had a good reputation, and it was convenient to reach from my home.

"You'll be living in an old war-time hut," the receptionist had told me. The huts were round-shaped, and sparsely furnished. When the course started, we were given lectures in the Ground School. The Ground School Instructors were matter-of-fact characters. "We'll teach you the rudiments of meteorology, principles of flight, Air Law, and technical aspects," the Chief Instructor had said. "If you complete the course successfully, you'll be awarded a Private Pilot's Licence."

When the students started flying, it was in the Tiger Moth aircraft, an open-cockpit bi-plane (i.e. two wings, one on top of the other). My flying Instructor, typical of the 1950s, was ex-military, and had a slightly jaundiced outlook on life. He had been instructing for many years, and appeared to have become weary of the scene.

In order to speak to the Instructor, we had to wear special leather helmets. In front of your face was an inviting Tube arrangement, called a 'Gosport Tube'. The idea was to yell into the Tube, and the Instructor could have heard you, and vice-versa.

"Put your goggles on, Biggles," a fellow student said to me one time.

"They don't fit properly," I complained.

"They look pretty cool, though!" he said.

The take-off in the Tiger Moth seemed complicated, because the aircraft had a tail-wheel. "As the speed increases during the initial part of the take-off run," said my Instructor, "the procedure is to push the stick forward to raise the tail-wheel off the ground." This seemed rather contrary to natural instinct. "Having raised the tail-wheel, allow the speed to pick-up. Then gently pull back on the stick. Easy does it!" At that point, hopefully, the Tiger Moth would have winged its way slowly into the air.

The sense of open-air flying was naturally exhilarating. It was a marvellous sensation flying along with the wind in your face, seemingly without a care in the world. The flat countryside of East Anglia stretched to the distant horizon, patterned with colourful fields. Life felt so good.

The Flying Instructor, however, had the job to teach the student on the first flight about the 'effects of controls'. "Try moving the control column from side to side," he said. That operated the ailerons, which caused the Tiger Moth to 'roll'. "Now move the control column forwards and backwards," said the Instructor. The elevators on the tailplane were moved, which made the Tiger Moth climb or descend.

Under each foot, the pilot had a rudder pedal. When pushed, the rudder pedal caused the Tiger Moth to 'yaw'. In order to turn the aircraft, the student was taught to apply aileron, and to control the tendency to 'yaw' by application of just the right amount of rudder.

"Now we'll have a go at some 'circuits and bumps'," said the Instructor. This was an exercise to teach the student how to complete a 'circuit' around the airfield, and then land on the grass runway at Marshall's. "The landing in a Tiger Moth takes a bit of getting used to, with its tail-wheel," said my Instructor. If you were really clever, a perfect three-point landing was made, with the main wheels and tail-wheel all touching the ground simultaneously. However, there was a good chance of a less than perfect three-point landing, in which case the Tiger Moth would have bounced back into the air. The Instructor would have grabbed the controls, at the same time as applying full power to the engine, and yelling abuse into the 'Gosport Tube'.

My Flying Instructor enjoyed doing aerobatics, and occasionally he demonstrated the techniques.

"We'll have a look at a 'slow roll'," my Instructor shouted. The 'Gosport Tube' did not always work efficiently. "Is your seat harness tight?"

"It seems fine," I said as I tugged at the straps.

"I'll start by slightly raising the nose of the aircraft." I followed him through on the controls.

"Now I'll move the ailerons, as if starting a turn. Feel the large amount of 'boot' I'm applying to the rudder pedals." I had noticed that already.

Cadets and Tiger Moth, 1959.

"That stops the aircraft from turning, and we are beginning to 'roll' around the horizon instead." The Tiger Moth was 'rolling' at a steady, controlled rate, and the aircraft was nearly upside-down.

"As we become inverted, note that we are having to push the nose up. We're changing from positive to negative gravity. The negative gravity can feel uncomfortable until you get used to it."

The Tiger Moth continued to roll through the inverted position, and gradually the gravity effect changed back again from negative to positive. Soon the aircraft was becoming level with the horizon once more.

"Now you have a go!" the Instructor had said. I nervously applied the aileron and rudder, as the Instructor had demonstrated, but the aircraft did not react in the same way. The nose dropped as we approached the inverted, and it felt nerve-racking as the Tiger Moth rolled jerkily and eccentrically when in the hands of an inexperienced pilot. By the time my effort had been completed, we had lost a considerable amount of height.

Navigation had its problems with the open-air cockpit. Once, on a solo cross-country navigation exercise, my map blew away on the first leg of the route. The carefully prepared map, with its lovingly drawn lines, and precise calculations neatly written down, wedged itself under one of the rudder pedals. I struggled unsuccessfully to dislodge the map, but nearly ended up in a spiral dive. "What a pain!" I thought to myself. "Either I'll have to return to base, and explain away the embarrassing predicament. Or I can just press on regardless, and hope to remember the route from all that careful map preparation."

I decided on the latter option. At first, it seemed to have worked. However, I unknowingly landed at the wrong airfield, a few miles from the planned one. After the landing, I innocently shut down the Tiger Moth's engine, and climbed out of the aircraft. I was hoping for a friendly welcome, and the opportunity to refuel. The place was strangely quiet, but eventually I did find somebody, and he was indeed quite friendly.

"Could I please arrange to refuel?" I asked.

"Sorry, mate!" he replied. "There's no fuel here."

The light slowly began to dawn.

After a few suitable pleasantries about what a nice airfield it was, I came to the dreaded but inevitable question.

"By the way, er, um, what was its name again?"

He confirmed the awful truth.

There was nothing for it, but to come clean, and to admit my error. Apart from anything else, the only way to have re-started the Tiger Moth engine was for someone to pull at the propeller, while I sat in the cockpit and moved the throttle.

Perhaps he was just keen to get rid of me. Whatever the reason, the brave man duly pulled at the Tiger Moth's propeller, and jumped out of the way as the engine burst into life (we had forgotten to put in any wheel chocks), and the aircraft leapt forward.

CHAPTER 4

Entering the RAF

The last landing I performed in the Tiger Moth, still clear in the memory, was immaculate. No bounces, no drifting, no nonsense; just smoothly and firmly placed on the right spot.

"I am pleased to present you with your Private Pilot's Licence," the Chief Flying Instructor had said. I must have left Marshall's of Cambridge with head held high, proud and confident. "Let's get out of here before I'm given the chance to mess up the next landing," I thought to myself.

However, the Tiger Moth had definitely whetted the appetite. When the Royal Air Force invited me to attend their Officer and Aircrew selection process, it seemed a natural progression.

The Wing Commander conducting the initial interviews was a stern character. He had a large bushy moustache, and appeared to be the stereotypical Air Force Officer. He noted that one of my 'A' level subjects was Economics.

"Perhaps you could tell me something about the 'Six and the Seven'?" he asked.

"Sir?" An awkward silence ensued.

"The Common Market and all that?"

"Oh! Ahem! Yes, of course. Well that bit isn't actually part of our course just yet."

"I see."

The aptitude testing was tricky too. The boxes, and squares, and triangular patterns which had to be re-adjusted, left the mind in a state of whirling confusion. However, at least I managed to complete the paper on time, which, we had been told was vital. When it came to the initiative and leadership tests, the arrangement of planks and ropes to cross a hypothetical river was really quite fun to do, and I was fortunately in a good team.

At the end of the two day selection process, when I had been numbered amongst the successful candidates, the Wing Commander drew me to one side.

"Congratulations!" he said. "Good stuff."

"Sorry about the 'Six and the Seven'," I said.

"When you get to that part of the syllabus at school," he replied. "I suggest you take careful note. That's the way this country is heading. The EEC is the thing of the future."

Before long, I was invited to become a Flight Cadet at the Royal Air Force College, Cranwell, in Lincolnshire, to join number 85 'Entry'. It was the year of the building of the Berlin Wall, 1961, when I appeared at Cranwell to begin my first term.

We were shown to our accommodation, a wooden hut shared by 4 new Flight Cadets. The floor covering of our huts was brown linoleum, kept in a highly polished state. In order not to have spoilt the polished effect, squares of soft rag were provided, and we had to slide around the linoleum, rather than walk on it.

"You can expect regular 'kit inspections'," our Drill Sergeant had said. "Lay out your shirts and other items on the bed, and make sure they're folded in the regulation way!" The Cadets spent hours preparing, and ironing their kit so that a good impression was given. They were taught also how to 'bull' the toe-caps of their standard-issue drill boots.

We were a diverse group. In our hut, there was Ted, a wiry lad with a strong north-country accent. At a public-speaking exercise, he gave an obscure talk about the benefits of eating haggis: "It went down like a lead balloon," he said.

"The haggis or the talk?" we asked.

"Both."

Eventually, the Staff deemed that he lacked 'Officer Qualities', and Ted left Cranwell.

"Hi you lot! I'm Chris." From the south of England, Chris had a cryptic sense of humour, and a reckless nature. He liked to be acknowledged as a triumph amongst men when it came to wooing the girls. He went on to complete the three year course at Cranwell successfully, helped considerably by the fact that he turned out to be a good pilot.

The third member of our hut was Brian, who came from the Midlands. Short and stocky, with a dry nature, he was a pipe-smoker, and had difficulty in coming to terms with the Cranwell 'system'. After a term at the College, he was invited to leave. Many years later, he entered the Guinness Book of Records by becoming the first person to fly solo in a micro-light around the world.

That left me as the fourth, and final member, of our hut. Tall and skinny still (not to mention spotty), my fair hair was cut even shorter. At least, along with Chris, I managed to survive the course at Cranwell.

To help with the domestics, each hut had a 'batman'. The one allocated to our hut was a demure Lincolnshire local, who took life very seriously. It was a strange relationship: he was middle-aged, and called the Cadets 'Sir'. Our Batman was obliging, and helpful as he fussed around doing not a lot, and at the end of each month it was made clear that he expected a 'tip'.

In spite of the spartan conditions and our differing back-grounds, or maybe because of them, each group in each hut quickly became firm friends. We spent hours chatting away; a favourite topic was inevitably the next three years to be spent at Cranwell, and the tedious prospect of the first year, devoted entirely to academics and ground subjects. If our Drill Sergeant, or a member of the 'Senior Entry', came into our hut, we were obliged to cease our conversation, and stand rigidly to attention.

When outside the huts, we had to wear some form of civilian hat, as uniforms had not been issued at that stage. The 'Entry' before ours had caused a stir when one of them, called Robin Hood, had been caught by the Commandant not wearing a hat.

"Good morning," said the Commandant to Robin Hood. "And who are you?"

The Commandant was a tall, imposing gentleman. His authoritarian air was complemented by a small moustache, neatly trimmed and immaculate. The Commandant was not wearing his military uniform that day, nevertheless his identity was somehow unmistakable. The Commandant wore a smart suit, with shiny shoes. Every part of the man was a flawless example to the Cadets.

Apparently, Robin Hood had not been over-hasty with his response to the Commandant's question. Robin Hood had decided to take his time before answering. That in itself had caused the Air Commodore a certain feeling of irritation.

Robin Hood had looked the Commandant firmly in the eye. At length he had drawled his reply: "I am Robin Hood," he had said. "And who are you?"

The Air Commodore's moustache had been set a-twitter by this reply. His eyes had glazed momentarily at the bald impertinence. He had rocked forwards and backwards slightly on his highly polished shoes. A faint colouring had appeared on his face for a few transient seconds.

Eventually, the Commandant had regained his composure. He then growled: "I'm Maid Marion. What do you think?" With that he had stormed off.

Our Drill Sergeant was a red-faced man with sandy hair, and keen blue eyes. He stood stiffly upright, and screamed and shouted at us. He also swore constant abuse at the noisy Jet Provost aircraft flying overhead (it would be another year before we got our hands on those). Behind the bluff and bluster was a good sense of humour, and he was in fact a fatherly figure to us. Every so often, the Drill Sergeant's parade ground voice softened, and he lapsed into the detail of some domestic drama involving his wife and daughter. It was a one-way conversation, though; any hint of our stepping out of line brought instant retribution.

One day the Drill Sergeant telephoned me.

"Pike?"

"Yes, Sergeant."

"What are you doing?"

"Nothing, Sergeant."

"Why not?"

"Er…"

"Never mind, never mind. Report to me at 1400 hours sharp!"

"Yes, Sergeant."

"By the way, Pike?"

"Yes, Sergeant?"

"I hope you are standing to attention while talking to me like this."

"Of course, Sergeant!"

The Sergeant had some tedious task in line for me, but it would not have crossed any of our minds to fail to turn up at his command.

After the first term, the Cadets were allowed to take cars to Cranwell. I had a bright red MG 'TC' which my father had given me. The MG had heavy steering, a knotchy gear-box, poor brakes, and no boot space. But it had a lot of personality.

Paddy, a good friend of mine, had an old-fashioned open-top Triumph. Paddy was the bain of the Drill Instructors: no matter how hard he (or they) tried, he could not avoid a somewhat dishevelled appearance. "The worst thing," said our Drill Instructor, "is that he doesn't seem to care!"

Paddy enjoyed racing around the countryside, with foot hard down, pontificating on some subject or other. Problems arose when it became dark, because his lights were unpredictable,

"Would you look at that?" said Paddy. "Only one headlight working again, and that's on full beam. These car makers nowadays, they haven't got a clue!"

It did not occur to him to have the car serviced, and on one occasion, when the one serviceable headlight failed, we ended up in a ditch. Undaunted, Paddy produced a torch, and we soon crawled back to Cranwell.

These character-forming activities took place in an atmosphere of huge amusement. However, the time raced by, and the end of our first year loomed. Next term, our flying training began.

CHAPTER 5

Pilot Training

The start of my Jet Provost flying was less than ideal. "The trouble is, he puts on a forced air of jollity," I said to fellow students as we discussed my Flying Instructor. "But this turns rather quickly to bad temper if things don't go well!" Underneath, the Instructor seemed a bit immature, and we got the feeling that he nursed an inferiority complex because he was not an ex-fighter pilot. When he went on leave, I was allocated another Instructor.

Suddenly, things began to click into place. The new Instructor, a tall, easy-going man, was a New Zealander, and had the knack of getting the best out of his students. Slowly but surely, the veils of mystery were pushed back, and I began to make progress towards the more advanced exercises, including night flying.

"OK gentlemen, the next phase of your course is night flying," the Flight Commander said as an introduction to his briefing. The Flight Commander was an intense man. He had a thick mat of black hair swept back from his forehead, and had a reputation for being slightly aloof. He proceeded to lecture us for several hours on the night flying procedures.

When the night flying began, there was tension in the air. "Flying in daylight is one thing, but at night it's a rather different ball-game," said my Instructor. "Just keep your wits about you, and there'll be no problem!" I felt sure that his fingers were firmly crossed.

We spent a lot of time practising night flying circuits. Sometimes 3 or 4 Jet Provost aircraft would be in the circuit together, and the importance of correct spacing between each aircraft had been emphasised during the briefings. On one of these circuits, I was half-way along the down-wind leg, when suddenly I became aware that I was being over-taken by another Jet Provost aircraft.

I nearly froze on the controls, and became aware of hairs on the back of my neck. The mind raced. How could it have been? Our long and exhaustive briefings most certainly had not covered this contingency.

The other Jet Provost was undoubtedly oblivious of my presence. "It was pure fluke that he didn't fly straight into the back your aircraft," my Instructor had said later, shaking his head.

"My instinctive reaction at the time was to press the toe-brakes," I admitted to him.

"Prat!" said my Instructor.

However, I managed to do something right: I throttled back the engine. By slowing down, at least some space was created between my aircraft and the intruder. His radio call to Air Traffic Control was naturally out of sequence, which caused some consternation. Eventually, he turned in, and landed ahead of me, and the crisis was solved. In the longer term it was solved another way: the pilot was deemed to be unsuitable, and became a navigator.

It was shortly after the night flying incident that another occurrence made disaster feel close. It was during a 'QGH' at Cranwell. The initials were just a neat way of telling Air Traffic Control that you were letting down through cloud in an agreed, and supposedly safe, procedure.

Half way through the 'QGH', there happened to be a momentary break in the cloud. At that precise moment, I looked up, and became aware of a Hastings transport aircraft lumbering through the area I was about to enter. That time, I did not bother with the toe brakes. I hauled the Jet Provost into a tight turn, applied full power to the engine, yelled at Air Traffic Control (and anybody else who happened to be listening), and climbed to a safe area to try again.

"A vital stepping stone is approaching," said the Squadron Commander one day. "The Basic Flying Test will be conducted by myself, or one of the Flight Commanders." It sounded terrifying, but in fact this man also had the ability to put his student at ease, or did in my case at least.

"OK," he said to me on the day of the Flying Test. "We'll start with some steep turns at height, and see how they go."

They went rather well.

He then asked me to perform stalls, spins, aerobatics, and circuit procedures which had been practised with my Instructor. These also seemed to go well. Suddenly it became apparent that I was actually quite enjoying a test flight; the concept had not even occurred to me. At the end of it, the Squadron Commander called me into his office. I noticed that he was hopping from one foot to the other.

"Well young Pike, what did you think of that then?"

I mumbled a bland reply.

"Well, I must say," he continued, "we had our doubts about you at first. But that went really well. Keep it up. Now push off. I'm dying for a pee."

Jet Provosts, Cranwell, 1962.

As I entered the crewroom, no doubt feeling about 10 feet tall, my colleagues had no need to ask how it went; they could read it on my face. For the first time, the goal I was seeking appeared to be a realistic one. It was an important personal moment. As my confidence began to get a toe-hold, I started to come out of my shell at last.

The next phase of training involved formation flying. I had a new Instructor, an ex-fighter pilot, who wore the dashing red and white chequered scarf of number 56 Squadron. A well-built, no-nonsense individual, he had a good sense of humour. He also had a certain way with words, not to mention women (he had a glamorous ex-model wife), and he was popular with the students. As a former fighter pilot, he had a lot of credence with the students.

"Right," my ex-fighter pilot Instructor said. "Just remember that this formation flying business is a lot of fun. And we're going to make you the best on the course!"

My new Instructor demonstrated the correct technique when flying in close formation on another aircraft. "The eyes should be kept roving," he said, "and not 'frozen' on the wing-tip, or any other part. Flying has to be as smooth as possible, anticipating the movements of the leader. Remember the throttle movements should be kept to a minimum. Resist the temptation of the learner to go from throttle closed to full throttle about ten times a second."

The Instructor made it look so easy, tucked in tightly next to the leader, following every movement flowingly. "These exercises are fun", said my Instructor, "but the techniques are just the same as those used for in-flight refuelling, and for other operational reasons."

As well as the formation flying exercises, we were taught the fundamentals of Instrument Flying. "The system of using the artificial horizon as the master instrument, scanning the other instruments, and then returning to the master instrument, should become second nature," my Instructor had said. "If the altimeter, for example, shows the height creeping up, check your artificial horizon before lowering the nose of the aircraft. Then re-check your artificial horizon. Your eyes must constantly scan the instruments. And remember that it's better to make small, regular adjustments than to allow a large error to build up."

When the formation flying, and instrument flying, phases had been completed, it was a sign that our time at Cranwell would be finished quite soon. For the final part of our Jet Provost course, the Flying Instructors introduced us to the rudiments of low flying, and low level navigation. "The public generally look upon low flying as an unnecessary, noisy, irritant to daily life!" said the briefing Officer. "From square one, therefore, we emphasise the importance of sticking to the minimum height of 250 feet. Also avoid built-up areas, danger areas, bird and wildlife sanctuaries, and sensitive areas. It's no easy task to navigate accurately at high speed and low level with all these restrictions, particularly with the fickle weather conditions we face."

Hours would be spent poring over maps, carefully drawing lines, and inserting tracks, times, distances, in little boxes on these maps.

By the early summer of 1964, the final handling tests for our 'Entry' were approaching. "If this flight goes well, you'll be deemed to have passed the course," the Squadron Commander had said. "You'll then be awarded with the coveted Royal Air Force 'Wings' at the Passing Out Parade." If the final handling test went badly, it was not unknown for the student to be asked to leave, even at that late stage. For the record, my final handling test was flown on the 12th of June 1964 in a Jet Provost Mark 4, and the flight lasted for 55 minutes. It was announced shortly afterwards that I had passed, and furthermore it had been decided to award me with a flying prize: the Dickson Trophy and Michael Hill memorial prize.

My father was asked by Cranwell to take the Passing Out Parade. My uncle was invited as well, so an unusual family reunion loomed.

"Well what do you know?" my friend Paddy had said. "Normally we have to put up with minor royalty, or a politician, or someone equally boring. But our 'Entry' are getting your Dad. Now that's what I call one-upmanship!"

Folland Gnats over Anglesey 1964.

CHAPTER 6

Gnats over Anglesey

The Passing Out Parade at Cranwell had been a moving and well-organised event. I had appeared at the rostrum to collect my pilot's 'wings' from my father, and then re-appeared later to be presented with the Dickson Trophy and Michael Hill memorial prize. As well as my parents, my uncle had attended, and a girlfriend, and everyone had been completely proud.

After the award of my wings, I had been posted to Valley for advanced flying training. It was the Autumn of 1964, the year in which Chairman Mao of China published his thoughts in a little Red Book. The Beatles were in full swing. It was the year after President Kennedy's assassination. It was the year of my 21st birthday.

Valley, on the Island of Anglesey, in North Wales, had the benefit of a micro-climate. Cumulus clouds sometimes built-up impressively over the distant mountains of Snowdonia. "It might be raining heavily in the mountains," the briefing Officer had said. "but here at Valley we often remain cloud-free in our micro-climate."

Shortly after arriving at Valley, I became indirectly involved in a bizarre accident. The 'crash alarm', which made an eerie noise, sounded over the Station Tannoy system. After a pregnant pause, a flustered voice came over the Tannoy: "*Crash… crash… crash. A Gnat aircraft has crashed on the airfield. All emergency services to rendezvous immediately.*" The stark announcement brought normal life to a momentary halt. It took a while for the mind to adjust to the new and unwelcome situation.

In the Station Medical Centre, where I happened to be at the time of the Tannoy message, the staff moved about quickly, preparing for any casualties. The Duty Doctor, a thin man with glasses and a studious air, was whisked away by the emergency ambulance. We wondered what form of gory scene the poor fellow might have to face.

It did not seem very long before the wailing ambulance returned. Two individuals were helped out of the ambulance, and ushered into the casualty reception room. I briefly glimpsed their pale faces, and torn flying suits. I recognised them both as Flying Instructors. I asked a nurse if I could speak to them, after their medical check-up. "We'll see," she said. "it may cheer them up to have a visitor. It's up to them."

It turned out that the Instructors were longing to speak to an aircrew visitor, albeit a humble student. They were dying to tell someone what had happened, and the Doctor had decreed that the authorities could not 'grill' them yet.

I entered the Ward nervously.

"Oh, it's you!" They sounded rather deflated.

"What happened?" I asked, directly to the point.

"Bloody hydraulic failure!" the Instructors said testily.

"Are you OK?" I asked.

"A few bumps and bruises," they said.

The Instructors went on to explain about the problem, which had involved an unfortunate quirk of the Folland Gnat: the procedure if an impending hydraulic failure was indicated. The tailplane had to be carefully 'frozen' within a restricted 'band', before all hydraulic pressure was lost. The tailplane then had movement over a limited range, using an electric motor. It was a weird and rather eccentric design.

The Instructors had been unable to 'freeze' the tailplane in the correct 'band'. Consequently, their attempt at landing had been too fast. The aircraft had touched the runway, and then bounced back into the air. At that stage, the Instructors began to lose control of their aircraft. They had little option but to pull their ejection seat handles.

Immediately after the two pilots had ejected, the Gnat had inverted, and buried itself in a sand dune by the runway. The pilots had landed in their parachutes close to the wreck of the Gnat.

After talking to the shaken Instructors, I had walked back to the Officers' Mess at Valley. The modern and utilitarian building was notable for an unpleasant odour of cheap antiseptic as the front door was entered. It was also our 'home' for the 6 month course at Valley. A number of students came up to ask about the accident victims. "They're not too bad," I had said, although the unspoken thought of most of the students was "if it can happen to two Instructors, what chance for us?"

In spite of its quirks, however, the Gnat was in reality an excellent aircraft for students. "It's handling is brilliant, it's so slick, and that rate of roll…!" one of my colleagues had said.

My Flying Instructor was an ex-Canberra bomber pilot. Fairly tall, with pale blond hair, and a red-faced complexion, he walked with a slight limp. Regrettably, he was not world famous for his good sense of humour. However, he enjoyed demonstrating death-defying turns, pulling right to the Gnat's structural limit of 7 'g' (7 times gravity force). Most aircraft were limited to 6 'g', which was usually the point when pilot's would have been 'blacking out' anyway.

After the demonstration, it was my turn to have a go.

"Go on, you bastard, keep pulling, keep pulling!" he said. I was used to 6 'g', but found it difficult to prevent myself from 'blacking out' at 7 'g', unless I was at the controls.

Then I noticed something.

My Instructor was also fine at 7 'g', but only if he was flying the aircraft himself. When I was at the controls, it soon became evident that his Instructor's patter was drying up at 6.5 'g'.

"Mary had a little lamb," I said at 6.5 'g'.

"Whaaaat you on aboooo.." I had increased to 7 'g'.

I relaxed the 'pull' back to about 6 'g', and asked: "Did you hear my question?"

"No. What question?"

"It's fleece was white as snow…" The 'pull' increased to 7 'g': the Instructor was silenced again. It became quite an art to develop this routine. After a while it almost felt like pulling puppet strings.

The low flying techniques started on the Jet Provost course were further developed, but as my Instructor put it: "It's a different scene altogether compared to the Jet Provost. Flying and navigating accurately at 420 knots and 250 feet is very hard to do properly. It takes years of experience to be any good." We flew over Wales, and over sparsely populated areas of Scotland.

"The speed of the Gnat means it's no problem to fly high level to Scotland, let down for the low level exercise, then climb back to height for the transit home," my Instructor had said. "The scenery along the west coast of Scotland will be dramatic on some of the routes we have designed for you," he said.

As well as the low flying exercises, the students developed their skills at formation flying. "If you want to know how to do it properly, just watch the 'Yellow Jacks!" my Instructor said. He referred to the aerobatics team of the Royal Air Force; it was before the days of the 'Red Arrows.' The 'Yellow Jacks' were based at Valley, and usually held their practise sessions late on spring afternoons, when student flying had ceased. The students normally stopped their map preparations, briefings, or other activities to gather outside and watch the 'Yellow Jack' performance with admiration. Set against the Snowdonia backdrop, the immaculate aerobatics routine held the fascinated attention of most of the students.

The 6 months course was at an advanced stage. "I'll warn you straightaway," said the Flight Commander, "the atmosphere around here will start to get a little tense from now on!" The Flight Commander, a bouncy character with flying experience on Hawker Hunters based in the Middle-East, frowned at us. The end of the course was near, and our future depended on the course results, and on the staff recommendations.

Rumours had been around that some of the Lightning squadrons were short of pilots, and that some of us were to be offered that option. "At the other end of the scale," said my Instructor, "is a posting to V-bombers. The thought gives most people the colliewobbles! Someone has to do it, of course, but many consider it to be the kiss of death." His words gave me nightmares.

As it turned out, the students on our course were very lucky: six of us, including me, were offered the Lightning.

We all felt on cloud nine.

CHAPTER 7

Hunters at Chivenor

"There's one more step before you lot get your hands on one of Her Majesty's Lightnings!" the Flight Commander at Valley had said.

Before our Lightning conversion course, six months had to be spent at Chivenor, in Devon, where we flew the Hawker Hunter ('Queen of the skies' to many), to consolidate flying skills.

"Your first couple of flights will be in the two-seat Hunter T7," the Briefing Officer at Chivenor had said. "After that you'll be flying the Hunter F6. For the first time in your lives, you'll be flying *real* fighters!" The attitude at Chivenor was different from previous units. The Instructors prided themselves in NOT being Qualified Flying Instructors; they were Weapons Instructors, and tended to look down upon 'Training Command'.

"Note the position of this score-board," the Chivenor Briefing Officer said as he showed us around. The board was in a prominent position by the Operations Room. "After each 'live firing' session on the range, your individual scores will be totted up. In the range, an aircraft, usually a Meteor, will tow a flag. Please note, the Meteor is not flown by remote control. It is towed by brave, underpaid pilots who will not appreciate any attempts at 'bravado'. Your Hunters will be loaded with live rounds, coloured for identification. After the Meteor has landed, the flag will be inspected. The number of holes, and relevant colour, will be noted and marked on this board." The students looked at each other, and thought: "this could be embarrassing."

In the live firing range, the students were taught to fly an accurate circuit, similar to the airfield circuit, but based on the towed flag rather than the runway.

"Fly initially about 1000 feet above the flag, following its heading," my Weapons Instructor had said. We were flying the two-seat Hunter T7, seated side-by-side, for my first session in the live firing range. The Weapons Instructor had a large black moustache. He was a Hunter pilot of the 'old school', with many flying hours under his belt.

"Your initial positioning, known as the 'perch', is pivotal to a successful exercise," said the Instructor. I flew an accurate heading, and moved our position forward slightly.

"When you're happy, call 'tipping in'."

I made the call.

"Now apply a large angle of bank, and turn towards the target."

As I turned, the flag began to loom up quite quickly.

"Reverse your turn at about this point," said the Instructor. "Standby to track the flag on your gun-sight." The gun-sight was a

Hunters and Meteors at Chivenor, 1965.

head-up display near the front windshield which showed a central dot, with small surrounding diamond shapes.

"The gun-sight display is rather twitchy," my Instructor said. "Don't be ham-fisted or it'll leap around all over the place." He was correct. "Careful now, or you'll never settle the gun-sight properly on the target." The gun-sight display was controlled by a gyroscope; it gave 'lead-lag', depending on the crossing angle and speed of the target.

The flag was rushing towards us. "Break left. Break left!" The Instructor suddenly called. I applied nearly 90 degrees of bank, and pulled to maximum 'g'.

"Return to the 'perch' position," said the Instructor.

"Sorry about that," I said.

"Don't worry, you're not the first. Things happen fast at the last moment, and you need to keep ahead of the game."

I adjusted our 'perch' position again, and looked across at my Instructor. "Try to relax", he said. "OK that looks a good 'perch', let's have another go."

I called 'tipping in', turned towards the flag, and then reversed my turn. As the flag drew near, I concentrated on the gun-sight.

"That's better," said the Instructor, "more accurate flying is the key in this case. Now concentrate on bringing the centre 'pip' on to the flag. Get ready to squeeze the trigger! Press the trigger… *now!*"

There was a loud *brrrrrrrr* and the whole aircraft shook as the Hunter's cannon were fired.

There was then no time to be lost. "Break hard left!" called my Instructor. "Watch out for getting mesmerised by the flag. As soon as you have finished firing, break away immediately."

As we returned to the 'perch', my Instructor said: "There's no way of knowing at this stage whether our cannon-rounds have hit the flag. We'll have to wait until the Meteor's landed."

We flew a few more circuits, then returned to Chivenor when our ammunition had been used up.

After landing, I waited apprehensively by the Operations Room scoreboard. When my Weapons Instructor appeared, he said: "Do you want the good news or the bad?" I raised my eyebrows.

"Well," he said. "That was reasonable for a first attempt. We scored 5 hits. That's the good news. The bad news is that the Meteor pilot wants to file a complaint against us for getting dangerously close to the flag. I've spoken to him, and I apologised. He said that he'll forget the complaint as it was your first session!"

I felt suitably humbled.

"Now you have had a few sessions against the flag," said Dave, our Briefing Officer, "we are going to look at the applied art! We're going to teach you the basics of air-to-air combat. The techniques we use have been developed from World War 2. To be any good you will need a mixture of cunning, skill, experience, brute force, and finesse. Oh! And a lot of luck, and a lot of courage." Dave had been a Weapons Instructor at Chivenor for a long time, and his speciality had become air-to-air combat.

"For your training exercises," the Briefing Officer continued, "two Hunters will take-off as a pair, and climb to 15,000 feet or so. You'll do the safety checks you've been taught, and then set up a 'circle of joy'." The 'circle of joy' meant that the two contestants would fly opposite each other in a circle, each pilot watching the other for the first sign of a move.

"The first move to break the 'circle of joy', " said Dave, "could be a number of options including, possibly, a 'yo-yo' manoeuvre. The intention will be to cut across the circle to gain advantage. Remember the aim of the exercise is achieve a position behind your opponent so that cannon-rounds (or other weapons) can be fired."

"During these gyrations, the two aircraft will lose height. It's essential that you do not bust your safety height. When you reach safety height, the exercise terminates, and you'll climb up for another practise."

After the extensive briefings, it was time to implement the theory.

I took-off as one of a pair of Hunters, and we set up our 'circle of joy', each aircraft seeking an opportunity to gain advantage. I suddenly recalled a hackneyed expression: "remember the Hun in the sun!"

The two Hunters were turning tightly in the 'circle-of-joy', and I decided to attempt a 'yo-yo' manoeuvre. My plan was to start the 'yo-

yo' just as I approached a position which put the sun behind the outline of my aircraft. At that point my 'foe' would lose sight of me temporarily as he was blinded by the sun.

I ensured that the Hunter's engine was at full power. Then I banked steeply towards my adversary, attempting to cut across the circle. The Hunter buffeted and rocked as I pulled 'g'. The aircraft lost height during the 'yo-yo'. My neck was strained hard round as I tried to keep visual contact with the other aircraft.

The manoeuvre worked partly.

The other pilot had indeed temporarily lost sight of me in the sun, which had caused him to slightly relax his turn. However, my 'yo-yo' had not been started soon enough, or flown sufficiently smoothly. The stalemate turning consequently continued to safety height.

We climbed up for further attempts, but eventually fuel shortage called a halt to our gyrations.

"That sounded OK," said Dave, as he summed up after our return to Chivenor. "You're learning! But give a thought to the real-life situation, with uneven types pitched against each other. There'd be no tidy training set-up. No briefed safety height. Just pilots desperately fighting for their lives, using every trick in the book." We agreed that even in training, there was an element of fear in our feelings.

CHAPTER 8

Lightning Conversion

"This is a red-letter day for you!" Flight Lieutenant Thorman said to me. He was a Flying Instructor with number 226 Operational Conversion Unit, based at Coltishall in Norfolk; my move to East Anglia had followed the completion of the course at Chivenor. "Today you'll experience your first flight in a Lightning aircraft. You'll probably remember it for the rest of your life."

Flight Lieutenant Thorman was not wrong.

The large size of the Lightning seemed awesome at first. The silver, slippery-looking aluminium fuselage housed two enormously powerful Rolls Royce engines, one on top of the other. To enter the cockpit, the pilot had to climb several steps on a specially designed ladder. At the top of the ladder, the pilot's first job was to ensure that safety pins were inserted at the top and bottom of his ejection seat. In front of the pilot was the typical instrument panel, with its Artificial Horizon, Air Speed Indicator, Compass, Engine Instruments, and various navigation systems. To the right was an unusual folding rubber tube. The pilot had to peer down this tube in order to see the picture produced by the Lightning's radar system, which was built into the nose of the aircraft. To the left would be found twin throttles, one for each engine. These were no ordinary throttles. When they were pushed fully forward, the engines would have been at full 'cold power'. Things did not stop there in the Lightning. When the twin throttles were rocked outwards, the engines went into 'reheat power'. That was a whole new ball-game. We had not come across anything remotely like that in our training so far.

The date was the 16th of August, 1965. I was scheduled to fly Lightning T4, a two-seater (with side-by-side seats), registration XM

997. My course (number 23) consisted of 10 pilots. We were in the middle of ground school training; today's flight was intended to whet the appetite, and to give the students a break from the ground school. We had already spent a few days at the School of Aviation Medicine, where the dire consequences of anoxia (oxygen starvation), and other medical nasties had been pointed out.

As I strapped into the Lightning, the carefully practised pre-start and other checks did not flow as well as hoped; perhaps I was a little nervous. There was an oxygen mask clamped to my face. There was an anti-'g' suit strapped round my waist, and down each leg. On my head was a white 'bone dome', which had built-in ear pieces to allow me to listen to radio calls, and to the Instructor. I wore a special flying suit, and cape leather flying gloves. I wore a life jacket on top of the flying suit. The whole paraphernalia felt bulky. However, it was not especially uncomfortable.

After the Instructor and I had strapped in, both ejection seat safety pins, for the top and bottom ejection handles, were removed. The perspex canopy was lowered by pressing an electrical switch, and the canopy was then mechanically locked.

Once both engines had been started, I called Air Traffic Control for clearance to taxi out to the runway. The Instructor controlled the Lightning for the first part of taxiing, but when a straight piece of taxiway presented itself, he gave me control. The disc-brake on each main wheel was operated by a vertical handle (like a bicycle brake handle) positioned on the stick by the pilot's hand. The nose wheel castored, so steering was achieved by gently pushing the rudder pedals left or right to produce differential braking. We soon approached the runway take-off point, and the Instructor took control again.

Air Traffic Control cleared us to line up on the runway, and suddenly this was it: the years of training and effort had somehow reached a climax.

For the Gnat and the Hunter, the Instructor had allowed me to do the first take-off. This time it was different. The Lightning Instructor held on to the controls.

There were some patches of cumulus cloud in the sky, and the wind made the day seem cool. Nevertheless, it felt hot in the cockpit as glimpses of sun bore down through the perspex canopy. Placed in the top of my white 'bone dome' was a darkened plastic shield, a form of in-built sun glasses, to protect the eyes. I made sure that the shield was lowered.

The Instructor glanced at me. Air Traffic Control had cleared us to take-off. The Instructor began to push both throttles forward. The engine noise increased markedly. The Instructor continued to manoeuvre the twin throttles towards the full 'cold power' position. I felt a rising sense of excitement and anticipation. Even with the well-padded 'bone dome' protecting the ears, the roar of the Rolls Royce engines was dramatic.

As the throttles reached full 'cold power', the Instructor released the brakes. I felt an immediate punch of acceleration in my back. The Air Speed Indicator started to move rapidly. The acceleration forces pushed me firmly into the seat. My mind found it hard to keep up with the speed and energy of the powerful machine. A jumble of emotions raced through my head, but overall I sensed an exhilaration as the runway dashed past at an ever-increasing pace.

Then the Instructor rocked both throttles outboard into the reheat position, and pushed the throttles as far forward as they would go. The effect was not instantaneous. There was a discernible fraction of time when the engines almost seemed to falter, as if drawing breath. This did not last long. A deep thunder was heard, and the whole aircraft was shaken, and hurled forward as if kicked by a giant foot. The Air Speed Indicator moved even more rapidly. My peripheral vision picked up a blur from the runway edges as we accelerated.

Suddenly, the nose wheel was raised off the ground, quickly followed by the main wheels. We were airborne.

The Lightning Instructor raised the nose to over 30 degrees nose up. It felt as if we were pointing vertically upwards. The acceleration forces continued to pin me to the seat. The altimeter looked a whirl as it span round in its efforts to keep up with the fantastic rate of climb.

My Instructor made clipped calls to Air Traffic Control, and we were cleared to Flight Level 360, which was about 36,000 feet. "The average height of the tropopause, the upper boundary of the troposphere," the Ground School Instructor had told us, "is around 36,000 feet. Above that height the air temperature is constant, or even inclined to increase slightly. The region of the tropopause is the most efficient height for the Lightning's Rolls Royce engines."

It would have taken a Hawker Hunter aircraft, for example, 10 or 15 minutes to reach that height. On that day, however, we decided to keep the engines in reheat and within two minutes or so, the Instructor was starting to level the Lightning at Flight Level 360. The rate of climb had seemed overwhelming. "Don't worry," said the Instructor, "you'll soon get used to it!"

The key to success, we had been told, was to think well ahead. If the mental processes got 'behind' the aircraft, then you were in trouble.

I was shown the aircraft radar picture, and how other aircraft appeared as a 'blip'. Later, we would be trained at how to interpret the 'blips', and how to manoeuvre our aircraft to a position where an airborne missile could be fired successfully.

Below, the outline of East Anglia, with the distinctive square shape of the Wash, appeared like a giant chart. We had no impression of speed at that height; the Air Speed Indicator had to be monitored, and it showed us to be flying at just below the speed of sound (Mach 1.0). "We'll do some turns, then have a look at supersonic flight. After that we'll descend to low level," said the Instructor.

The Gnat and the Hunter aircraft had been capable of supersonic flight, but it had been a struggle. Both types of aircraft had to be climbed as high as possible, dived out to sea with the engine at full power, then, after a quick foray into supersonic flight (with the Air Speed Indicator showing just above Mach1.0), it had been necessary to recover smartly because of the rapid height loss.

"You have control," said the Lightning Instructor. For a second I thought "Who? Me?" However, the Lightning, with its hydraulically assisted flying controls, was light and easy to fly.

"Try a few turns." They were no problem at all.

"Great. Now push the throttles to full 'cold power'." The aircraft reacted; the Air Speed Indicator hovered just below Mach 1.0. "Don't descend." We were pointing away from the Norfolk shore, and the coastline of the Low Countries, on the other side of the North Sea, was faintly visible in the far distance. "Now engage reheat on both engines, just as you have practised in the simulator." I made sure that both throttles were in the fully forward 'cold power' position, and then nudged them both outwards. There was a fractional hesitation, similar to the take off, and then *whoosh!* The aircraft was thrown forward. The Air Speed Indicator jumped above Mach 1.0; we were flying at supersonic speed without any need to descend, indeed, without any fuss at all.

"Now push the throttles forward from minimum to maximum reheat position." As I did so, the Air Speed Indicator began to increase rapidly: Mach 1.1, Mach 1.2, Mach 1.3.

"OK. Hold it there." I eased back the twin throttles.

"Now try a turn." I gingerly applied some bank. "Go on! It won't bite you." I applied 60 or 70 degrees of bank.

"'That's better. Now pull some 'g', but maintain Mach 1.3." I pulled back hard on the stick, but unlike subsonic flight, it proved impossible to pull to the point where the aircraft started to buffet. At supersonic speeds the airflow over the wings was infinitely smooth, and would not 'break away' and cause buffet. We had learned the theory in Ground School. This was the living proof.

"Maintain your speed in the turn." I eased the throttles forward to hold the airspeed at Mach 1.3.

"Well done. Now, keep your turn going, and ease the throttles out of reheat." Immediately the speed started to reduce. As soon as Mach 1.0 was reached, the aircraft started to be buffeted and knocked around.

"Right," said the Lightning Instructor, "what's happening to the fuel?" A quick inspection showed that we had already used over half the fuel available.

The Instructor made a radio call, and told me to turn towards Coltishall, at the same time as starting a descent.

"Don't forget to use the airbrakes." A switch on one of the throttles activated a hydraulic pump to push out two board-like surfaces at the rear of the Lightning. Immediately, we felt the effect as the aircraft started to slow down, and I therefore lowered the nose to maintain our airspeed, and to increase the rate of descent. We were closing rapidly on the Norfolk coastline. Soon, the altimeter showed us to be approaching 5,000 feet.

"I have control," and the Instructor took hold of the Lightning's flying controls. "I'll demonstrate a quick 'loop'," he said.

We ensured the area was clear of any other aircraft, then the Instructor straightened the wings, and pulled the Lightning vertically upwards. 6'g' appeared on the 'g' meter, and the bladders in my anti-'g' suit inflated. The bladders, by squeezing against the legs and pelvic area, prevented blood from pooling in the lower limbs. This in turn helped to counter the danger of the pilot 'blacking out'. A variable valve influenced the anti-'g' suit; the higher the level of 'g', the more the suit would squeeze. As we maintained 6'g' for the start of the 'loop', I felt the powerful pressure against my limbs when the suit inflated to maximum.

The Instructor pushed the throttles forward to full 'cold power', and we began to climb rapidly through several thousand feet. At the top of the 'loop', the altimeter stopped racing upwards, and soon we could see the sea and – in the distance – the Norfolk coastline appearing in the top of our canopy. Smoothly, he continued to fly the Lightning down the other side of the 'loop.'

"That's enough of that," he said on completion of the 'loop', "we'll have a look at the circuit procedure, and then we'll have to land. We're already quite low on fuel."

I took the controls again, and headed towards Coltishall. We made some more radio calls, and soon Coltishall airfield was racing towards us. "Use the airbrakes to slow down." Sure enough, a flick of the switch, and the airbrakes had a dramatic effect. I brought the speed back to below 200 knots, and reduced our height to 1000 feet. The Lightning Instructor took the flying controls, and started the 'downwind' leg of the circuit. He operated a switch to lower the undercarriage as part of the 'downwind' checks, and another switch lowered the 'flaps'.

As we drew level with the runway touchdown point, he applied about 60 degrees angle of bank and turned the aircraft towards the runway. During this 'finals' turn, we reduced height, and as we rolled out on 'finals' – pointing at the runway – the Instructor aimed for the runway edge. The Lightning landed fast; we needed all the runway we could get. Just as the wheels were about to touch down, the Instructor pushed the throttles to full 'cold power.'

"Time for one more," he said. He flew the circuit pattern again, but next time at the touch down point, he placed the Lightning wheels firmly on the runway, close to the runway edge.

"These tyres are specially made," he said, "but even so will cope with only 10 landings on average. If there's a strong cross-wind, just one landing will finish them." The Instructor operated another switch, and we felt a rapid deceleration as the landing parachute at the back of the Lightning billowed out of its housing.

"OK. Taxi me back to the dispersal area." As we nosed our way past other parked Lightnings, an Airman marshalled us to the allocated parking spot. I applied the parking brake, and went through the shutdown procedure.

"Well done," said the Instructor as we climbed out of the cockpit. "Enjoy it?"

"Too right!" The answer was characteristically low-key, but inside my mind was a thrilling turmoil of excitement.

We continued to chat as we walked towards the engineering 'line hut' to make entries in the aircraft log. My sense of enthusiasm lingered. Before he signed the document, my Instructor wrote down the take-off and landing times in the log.

The flight had lasted exactly 35 minutes.

CHAPTER 9

Radar Sorties

"We've got a tough task!" said John, one of the Lightning Instructors at Coltishall. Slightly balding, and somewhat assiduous by nature, he was hardly the archetypal fighter pilot. "Most of you'll have your hands full just coping with flying the Lightning," he said. "But I'm afraid we'll be asking you to do a bit more than that." Looking rather serious, he went on to explain the basic principles behind interception profiles.

On the left side of the Lightning cockpit, just above the twin throttles, the pilot had a hand-grip. This was used to control the aircraft radar and weapon system. "There are over a dozen different functions on the hand-grip," said John. "Once you've learnt them all, you won't need to look down. Everything will be done by feel."

To fly a successful interception, the Lightning pilot talked to ground-based fighter controllers sitting deep under-ground, watching radar screens. "Your own radar screen in the Lightning is tiny in comparison, and it has its limitations," said John. "The fighter controller, therefore, will aim to get you to a position where a target will show up on your radar. From then on, you'll control the interception."

In October 1965, I flew my first 'radar' sortie, with Terry, who was a weapons Instructor. Terry had the ability to combine academic theory with day-to-day practicality in a convincing way.

"We'll start with targets crossing at 90 degrees," said Terry. "It's the easiest interception type to interpret on the radar screen. Once the target has crossed the nose, we just turn in behind it, making adjustments during the turn."

I had levelled the Lightning at 36,000 feet, and we were talking to the radar controller at Neatishead. We had taken-off from Coltishall as number two in a pair of Lightning aircraft.

"Your playmate is on the left, range 35 nautical miles," said Neatishead.

"Roger. Looking," I replied. I searched vigorously with the hand-grip controller. We had practised in the simulator, and had learnt the 'feel' of the hand-grip.

"Steady down the movements," said Terry. "You'll never pick up the target like that." On the side of the radar screen was a small line. This showed the angle of elevation of the radar dish, situated in the nose of the aircraft. "We know the target is above us, so search in small movements just above the 'level' mark."

At around 30 nautical miles on the radar scale, a faint 'blip' was appearing.

"That looks like it," said Terry. "Just hold this heading for now."

"Neatishead, Lightning 996, Tallyho," I called.

"Roger 996," replied Neatishead.

"As we track the target," said Terry, "we have to assess its angle of crossing. This one, at 90 degrees, isn't too bad. When the target is at an angle of, say, 150 degrees, it's more problematic to assess. As you see the 'blip' moving down the radar screen, try to get a mental picture of what it's actually doing outside."

The target was approaching the 20 mile point, and was moving slightly left to right on our screen.

"If the 'blip' maintains a constant position all the way to the bottom of the screen," said Terry, "it means we're on a collision course."

However, the left to right movement of the 'blip' became more pronounced.

"I reckon this one's going to pass 6 or 7 miles ahead," said Terry. "It's a bit too far, so we'd better cut the corner. Turn right 10 degrees."

I made the turn. As the 'blip' passed through 12 miles, Terry said: "Turn right another 10 degrees." The 'blip' continued its relentless march down our radar tube.

"Try to 'lock-on' to the target," said Terry. I moved the hand-controller to position a circle over the 'blip', and pressed a switch on the front of the hand-controller. It had no effect. "Try again," said Terry. Next time I pressed the switch, the radar picture changed. "That's better," said Terry. "The radar has 'locked-on', and will automatically track the target now."

At a range of 4 miles, the 'blip' was crossing our nose, and I briefly looked out of the cockpit and saw the outline of the other Lightning.

"No peeping!" said Terry. "This is supposed to be done purely on radar, simulating operating in cloud."

As the 'blip' passed our nose, Terry said: "OK. Start you turn on to target heading now."

The radar continued to track the target as we turned in behind it. "We're going to roll out just over two miles behind," said Terry. "Continue closing on the target, and we'll do a 'visident'." Terry called out the range as we got closer to visually identify (visident) the target. Then he said: "OK turn right to put the radar 'blip' to one side."

Our radar showed the target to be well over on our left, and we were slightly below it. "Look up now," said Terry. As I looked up, the other Lightning appeared to be alarmingly close. I could clearly see the other pilot, and I could read the hull-letters of the aircraft.

"Well done," said Terry. "Now break away to the right, and when we're safely clear, ask Neatishead to set us up for another one."

"It'll take months of practice," said Terry later, when we de-briefed the flight. "All we can do here is show you the basics."

Towards the end of the course at Coltishall, we were presented with alternatives for our choice of Squadron. There was the possibility of Scotland, or Binbrook in Lincolnshire, for a lucky few Germany was an option, but my first choice was Wattisham in Suffolk. The latter meant a posting to number 56 Squadron, which by chance had been my father's first Squadron. Also at that stage, I seemed to be getting on well with the new Commanding Officer, a fellow member of my course.

When my posting to 56 Squadron was confirmed, I felt thrilled, and in January 1966 moved from Norfolk to Suffolk. The feeling of elation was quickly dampened.

One of my first duties was to attend the funeral of a young Lightning pilot of 56 Squadron. The accident happened shortly before my drive south.

The pilot had been unable to fully lower his aircraft's landing-wheels. One of the legs of the spindly-looking Lightning undercarriage had jammed half-way. The emergency flip-cards gave instructions which included pushing negative gravity, pulling positive gravity, and various other manoeuvres, all in an attempt to free the stuck leg. If these actions failed, the flip-card instructions were unequivocal: "Fly the aircraft out to sea (if possible), and eject. *Do not attempt to land."*

As the pilot crossed the coast, he levelled the aircraft, and engaged the auto-pilot with the Lightning pointing out to sea. He then raised both hands to the top handle of his ejection seat, braced himself for the ejection, and pulled the handle.

Nothing happened.

The pilot's actions that followed were surmised by the Accident Investigation Board convened after the accident. The Board conjectured that the pilot tried to pull the top handle again, but still nothing happened. The pilot therefore attempted to pull the alternative ejection-seat handle on the bottom of the seat, but with the same result.

Air Traffic Control listened with anguish to the pilot's radio calls. The Duty Pilot ascertained that the aircraft canopy had not jettisoned as intended, and therefore the ejection sequence was interrupted. The pilot was desperately short of fuel by then, but he elected, as a last resort, to return to Wattisham in order to attempt a crash landing on the runway. The emergency services at Wattisham were brought to a high state of readiness, and back-up civilian resources were rushed to the airfield.

The Lightning ran out of fuel just a few miles short of the runway. The aircraft landed in a field, and the Doctor reckoned that the deceleration forces on touch-down killed the pilot instantly.

The funeral service was held with full military honours, which included the firing of volleys. The service was in a local Church, fine and timeless in its country setting. The bereaved family, brave and reserved up to the time of the volleys, broke down in sobs when the rifle-firing began.

It was a sombre and tragic start to my career on 56 Squadron.

Another unexpected aspect gave a negative spin to my new Squadron. The in-coming Commanding Officer, my Coltishall course colleague, went through an unfortunate change. Gone was the carefree attitude, and instead we had a tense and worried man. His number two was a dour, scheming Scotsman with his own ambitious agenda. The pair became known as Batman and Robin. "The harsh truth is that they're just not up to the job," said some of the old hands.

The year of 1966 was consequently over-shadowed by the gloomy atmosphere caused by the fatal accident, and by the Batman and Robin syndrome. Nevertheless, in spite of the problems, the excitement and challenge of my first Lightning Squadron was exhilarating.

In the early months of the following year, a local attraction appeared on the Mess notice-board one day: those interested were invited to a Suffolk Barn Dance taking place in the neighbourhood. Five of us decided to share a lift to the Barn Dance.

After driving through the attractive countryside, we parked, and walked to a large marquee where a Bar had been set up. Young people were arriving all the time, but one in particular caught our attention as she got out of her car. Dark, well-groomed hair complemented a curvaceous figure in a cat-suit. Someone made a comment. Then she disappeared.

A local Band started, and a few people danced. The evening wore on, slightly plodding, but our group was determined to make the best of it. When it was time for the Band to have a break, a disco was started. It was the swinging sixties; the Top Ten chart was crowded with spectacular music, ideal for dancing.

Someone dropped something. I leant forward to pick it up. "Thanks," she said, sounding surprised. I blinked. It was the cat-suit babe. The light was poor, but our eyes met briefly. Then she was whisked away to dance with someone. She disappeared again.

However, I could see the dance floor, and after twenty minutes or so, she appeared by the bar. I plucked up courage. "Would you like to dance?"

She turned to me, and looked at me critically.

"Thank you," she said lightly.

It was a strange feeling. The cat suit icon, remote and untouchable as we ogled her arrival, was at that moment dancing with me. Furthermore, she seemed to be quite enjoying it. As we twisted and danced to the groovy music, we looked at each other. She had an enquiring, intelligent face; a ready, sweet smile.

The evening wore on and, unusually, I was holding back the others who were piling into the car for the drive back to Wattisham.

"Your friends are getting impatient," she said. I shrugged.

Then I summoned-up more courage. "Is there any chance we could meet again?" I asked. There was a pause. *"Up…up… and away, in my beautiful balloon…"* sang out from the Disco. We heard a car horn. I felt awkward. I also felt another critical look in my direction. "You'd better go now, or your friends will leave you behind."

"Sure. Thanks for the dancing. It was great."

"Richard?" Her eyes, kindly and alive, searched for mine.

"Will you ring me?"

The Author in the cockpit of a Lightning F3.

CHAPTER 10

Off to Pisa

As we drove back to Wattisham, I had a feeling of excitement.

"Batman and Robin won't like it," said one of my companions. "They'll have you arrested for enjoying yourself."

"Batman and Robin suck!" said someone.

The driver swerved as everyone burst out laughing. The twisting Suffolk lanes proved challenging at that midnight hour.

"Oh no!" groaned the driver. "Batman and Robin will go bananas when Taceval starts."

There were a few more moans from the car passengers as the prospect of a forthcoming Tactical Evaluation was discussed. A Team was expected to visit Wattisham at any time, and Batman and Robin would have felt their careers on the line.

When the Team did turn up, it happened that I was on duty in the Quick Reaction Alert hangar. In the hangar, two Lightnings were at readiness to be airborne within ten minutes of a siren sounding. Often, it was dull work, and the crews would sit for hours awaiting action.

Suddenly, an Officer we did not recognise walked in through the door, and placed himself in the small operations room. He did not say a word. The two pilots on duty looked at each other. They tried to make polite conversation, but the silent one was not interested.

Then the Alert siren sounded. Before climbing into their aircraft, the two pilots rushed to pick up life-jackets. Calamity stared them in the face. There was only one life-jacket. The other pilot had forgotten to bring his. In the scramble, he took the only life-jacket available: mine. I climbed aboard my Lightning anyway, life-jacketless. The silent one made copious notes. We felt embarrassed.

Fortunately, we were not required to get airborne, and a spare lifejacket was hastily sent over. The other pilot apologised. Batman and Robin, as predicted, went bananas. The silent one made even more copious notes.

"It was a bad start," Batman correctly pointed out.

"The Taceval Team were not impressed," Robin also correctly pointed out.

"Don't do it again," said Batman.

"Or you can't go to Cyprus," said Robin.

"Cyprus?" we asked.

"Cyprus," said Batman and Robin in unison.

We were then told about plans which were afoot to send a detachment of four Lightnings from 56 Squadron to Akrotiri, in Cyprus, staying there for a month. The aircraft would fly out non-stop, using Victor Tankers for in-flight refuelling. "You'll have extra training sessions to improve your in-flight refuelling techniques," said Batman.

The Cyprus detachment was planned for the Spring of 1967, the year of the Arab-Israeli Six-Day War.

"The Victor Tanker has two pods, one under each wing," said the Briefing Officer. "The Tanker trails refuelling pipes from these pods. At the end of the pipes is a basket arrangement which has a ring of green lights for night-time 'Tanking'. Your job is to fly accurate formation so that you can insert the refuelling probe on the side of your aircraft into the Victor's basket. Once good contact has been made, a spring in the basket will grip the probe…"

"Painful," said someone.

"…and fuel will transfer from the Victor's tanks to yours."

It sounded good, but in reality it was not that straightforward. We found that as the Lightning pilot approached the basket, the airflow around his aircraft pushed the basket away. There was sometimes turbulence, which caused the basket to bounce around, and inevitably the efforts of a pilot would become frenetic if he was low on fuel, and kept missing the basket.

"A weak point has been designed into the probe," said the Briefing Officer. "If everything goes completely pear-shaped, the probe will break off. You'll then be forced to land at the nearest diversion airfield."

As promised by Batman, some hours were spent practising in-flight refuelling before the start of the mammoth flight to Akrotiri.

However, the day of departure, 11 April 1967, soon dawned. Last minute plans were checked and double-checked. Our call-sign was 'Phoenix' (a Phoenix was part of the Squadron emblem).

"Phoenix detachment to cockpit readiness," Bawdsey eventually called: the radar unit had to time our take-off to allow a neat interception of the Victor Tankers as they flew south from their base at Marham, in Norfolk. All four Lightning pilots strapped into their aircraft, and sat, waiting for further instructions.

"Phoenix 10 and 11 scramble!" ordered Bawdsey. The other two aircraft started-up.

I was in the second pair of Lightnings, with Roger as the lead aircraft of our pair. A keen and perceptive individual, Roger eventually became an Air Marshal.

Soon it was our turn. "Phoenix 12 and 13 scramble!" called Bawdsey.

I glanced across at Roger after pressing the two Rolls Royce engine starter buttons. Ground-crew moved around checking our aircraft during the start sequence. There were no signs of fire, both aircraft seemed fine, and Roger gave me a thumbs-up sign. Soon, we were taxiing on to the runway for take-off.

During the climb-out from Wattisham, I was tucked into a tight close formation position on Roger as we flew through layers of thick cloud. We were bounced around by turbulence, but before long we levelled at altitude, just above the copious layers of cloud.

"Phoenix 12 and 13, your Tankers are range 45 miles, just left of the nose," said Bawdsey.

"Roger," said Roger.

The visibility was reasonable at that point, so I eased out from my close formation position into a 'loose echelon'.

"Tankers now range 30 miles," said Bawdsey. I was searching on my radar, and detected three distinct 'blips' as Roger called: "Tallyho."

Bawdsey continued to count-down the range of the Tankers, and soon the characteristic outline of the one-time V-bombers could be seen visually.

"You're clear for re-fuelling," called the Captain of the lead Tanker. He was prudently checking the re-fuelling systems at an early stage. When they had been proved, without problem, Roger and I eased out into comfortable loose-formation positions, one Lightning on each side of the Tanker.

A flight time of about 5 hours was anticipated to Akrotiri, and the Officers' Mess had supplied us with packed lunches. After several in-flight refuels, and as the formation approached the northern part of Italy, Roger decided it was time to feed the inner-man. He examined the contents of his white-cardboard lunch-box. Inside, he found an unexceptional collection of boiled eggs, ham, salt and pepper, bread rolls and some fruit.

"I'll do this properly," thought Roger, and he began to lay his lunch neatly around various parts of the cockpit. The bread rolls were placed near the twin throttles. Boiled eggs rested in a convenient fold of the radar cover. Fruit was positioned on the top coaming, together with salt and pepper. Meanwhile, Roger ensured that the Lightning's auto-pilot was engaged so that he could concentrate on the lunch.

"Phoenix 12 and 13, this is the Tanker Captain. Be advised a band of turbulence is showing on our radar. Suggest you move in closer." It was not unusual. The northern Alps often caused in-flight turbulence, as most airline pilots would testify.

By the time Roger had grabbed the controls, dis-engaged the Lightning's auto-pilot, and attempted to catch the boiled eggs, he had already trodden on the bread rolls, and combined the fruit with the salt and pepper on the cockpit floor.

"Phoenix 12 and 13 from Tanker Captain. You are clear in to refuel now. We'll start this one earlier than planned in view of the turbulence."

I moved to the familiar 'hold' position a few meters away from the re-fuelling basket, and waited. The basket bounced up and down disturbingly. I moved a little bit closer, and the basket was pushed away by the Lightning's airflow. "Relax," I told myself. It was

Lightnings in-flight refuelling 1965.

The Author wearing flying headgear.

important to hold a formation position on the Tanker itself, rather than to chase the basket. I eased forward some more, hoping to choose a moment when the basket was relatively steady.

To help avoid throttle over-control, I moved just one of the Lightning's twin throttles, leaving the second one in a constant position. At an opportune moment, I pushed the throttle forward. It did not work. I had missed the basket, which banged itself angrily against the canopy. I eased back the throttles to re-position for another attempt. I glanced across at Roger, and noticed he was having difficulty as well.

As I manoeuvred back to the 'hold', I noticed that Roger was moving away from the normal position. "Phoenix 12 from 13, got a problem?" I asked.

"Affirm," said Roger. "Standby." I could see nothing obvious, but moved back from the basket, and waited.

"Phoenix 13 from 12, my re-fuelling probe has broken off," said Roger. "Tanker Captain from Phoenix 12, confirm nearest diversion, please."

"Standby Phoenix 12," said the Tanker Captain. Then he called: "Phoenix 12 and 13, your nearest diversion airfield is Pisa. Suggest initial heading of 195 degrees. Standby for further information."

"Wow!" I thought. "A night in Italy. Can't be bad." I closed up to the other Lightning, and tucked into 'echelon starboard' for the let-down through cloud. The Tanker Captain gave us the radio frequency for Pisa, and we bade him a premature farewell.

Roger called me over to the new radio frequency, and then made a transmission: "Pisa Approach, this is Phoenix 12 and 13, do you read?" There was a strange noise, a scratching sound, followed by an Italian voice on the radio.

"Eee. Who a-calla Pisa?"

"Pisa this is Phoenix 12 and 13."

"Felix 12 and 13. What you wan, huh?"

"Phoenix 12 and 13, request diversion to your airfield."

There was a pause, and occasional flustered sounds would drift across on the radio.

"Felix, we no expect you. Go away, huh?"

Roger attempted to summarise our situation to the Pisa controller.

"Felix. Pis-a off. Go away. You no declare a emergency, huh?"

Roger must have thought: "That's easily done." The next thing we heard was: "Pan. Pan. Pan. Two Lightning aircraft short of fuel, request immediate diversion to Pisa."

"Felix. You declare emergency, huh?"

Roger repeated the emergency call in full.

"OK OK," said the Pisa controller, and he gave us an adjustment to our heading. The controller seemed to resign himself to our diversion, and we elected to have a Ground Controlled Approach from radar.

Soon, we touched down on the runway at Pisa, still in close formation.

We closed down the Rolls Royce engines, and climbed out of our aircraft. A brightly coloured van from Air Traffic Control had stopped by Roger's aircraft, and the driver appeared to be gesticulating as I wandered over. Roger ignored him, and inspected the broken probe on his Lightning.

As I got closer, Roger looked at me fiercely. "I'm starving," he said.

"Me too," I replied.

"Tell this guy to Pis-a off and get our lunch will you?"

The Air Traffic Control Officer stood gaping as two Lightning pilot's burst out laughing.

CHAPTER 11

Fun & Farewells

A few days were needed before our flight to Cyprus could be concluded.

"We'd better see the Leaning Tower while we're here," said Roger. "And some other local attractions." He winked at a girl walking by.

At length, a rescue party arrived in a Hercules Transport aircraft. Engineers fitted a new refuelling probe to Lightning XP 707, and we carried out functional tests. Eventually, the repaired aircraft was cleared to continue its flight to Cyprus. The two Lightnings consequently took-off in close formation, and we bade farewell to the Air Traffic Control Officer at Pisa. "Good-a-bye," he said. "Have a good-a trip." We then changed radio frequency to speak to the radar controller, and to the Victor Tanker. After a rendezvous with the Tanker, our flight to Cyprus was resumed.

It was fairly late in the afternoon when we eventually touched down at Akrotiri, in the Sovereign Base Area on the southern tip of Cyprus. As we climbed out of our Lightnings, Roger commented on the heady, sweet smell of orange blossom. There was a buzz of activity all around; Akrotiri were coping with a large number of visiting aircraft.

"Made it at last," said the other Lightning pilots, as they walked over to greet us. "No time to be lost. Holiday's over. Briefings start in ten minutes."

"We've organised some practice air combat training," said the Detachment Commander as he began the briefing session. "The fine weather here means we can safely reduce our minimum landing fuel. It'll give a bit more time for the air combat." The briefings then continued for what seemed like hours.

The next day, we had an early start to commence air combat training before temperatures became excessively hot. "The Lightning is a surprisingly forgiving aircraft in air combat," the Detachment Commander had commented during the previous night's briefings. "If you end up in the classic worst case, pointing vertically upwards with the airspeed reducing to zero, all is not lost. Just centralise the flying controls, position the throttles at about half cold-power, and wait."

We had opportunities to apply the advice. As the aircraft weaved and turned in the clear and uncrowded Cypriot skies, the crews progressively improved their air combat skills. Nevertheless, they occasionally ended up in the classic scenario: 'Nose vertically up, and nothing on the clock.' It took a few moments, but eventually the Lightning would fall cleanly forwards, backwards or side-ways. When the air-speed built up again, the aircraft would be back in the fray rapidly.

"It can't be all fun while we're here," said the Detachment Commander as he greeted us after an air combat session. "We've been asked to hold one aircraft on 'Quick Reaction Alert'."

Arrangements were made for the pilot on 'QRA' duty to stay in the vicinity of the Operations Room, waiting for the call to 'scramble'. A Land Rover was provided at the back of the area for the pilot to drive himself the short distance to the parked Lightning.

Early one morning, I was on 'QRA' duty, when an order to 'scramble' was received. It was still quite dark, and the milkman was making his dawn delivery to the Operations set-up. His van blocked the Land Rover.

"Good morning, Sir," said the milkman cheerfully. "How are we today?"

"Fine thanks, but…"

"The usual number today, Sir?"

"Fine thanks, but…"

"You going somewhere, Sir?" he asked.

"Well…oh never mind." I abandoned ideas to use the Land Rover, and sprinted across to the waiting Lightning. I climbed into the aircraft cockpit, panting.

"Radar this is QRA checking in," I called breathlessly.
"Thought you'd got lost!" said Radar.
"The milkman nearly hi-jacked me," I replied.
"Roger," said Radar, "Head 360 degrees for possible target, *scramble, scramble, scramble.*"

As I climbed away from Akrotiri, the distinctive shape of the Island of Cyprus appeared below. The dawn colours were just emerging. The sky was cloudless, and a strange agitation pervaded the atmosphere. It was the 18th of April 1967; the Turkish invasion of the Island was still some eight years away. The United Nations Peace-Keeping Force had been on the Island for three years. Makarios was in power, and there was a restless feeling. I over-flew Nicosia on my northerly heading, and then Kyrenia on the far coast. Soon, I was flying over sea again, maintaining a direction towards the Turkish mainland.

"Your target has faded," said Radar. "Turn left onto 270 degrees initially, and set-up an east-west Combat Air Patrol." To my left, I could see Kyrenia in the distance. north of me, to the right, the mainland of Turkey was barely visible.

I had spent around 20 minutes in the Combat Air Patrol pattern, when Radar called: "QRA confirm fuel state."

"Five more minutes on patrol available," I replied.

"Roger. In that case return to base now. I have no targets showing."

As I turned on to a southerly heading, and crossed the Cypriot coast again, there were signs of the population coming to life at the start of the day. At length, I called Akrotiri on the radio.

"You are clear to land," said Akrotiri. "No conflicting traffic."

"I should hope not at this hour," I replied.

In spite of the hour, the Detachment Commander was around after my landing.

"How did it go?" he asked.

I explained about the Combat Air Patrol.

"By the way," he said. "I've got some news for you. Do you want it now, or after breakfast?"

"Go on," I said.

"The rumour of a permanent Squadron posting here has been confirmed. A new Squadron is to be formed at Wattisham to take our place. Batman has offered you the carrot of an Instrument Rating Examiner's course if you volunteer for the new Squadron."

"I think I'd better sleep on it," I replied. "But I must admit it sounds interesting."

❖ ❖ ❖

Things moved fast in the Lightning world, and so it came about that the following month my allegiance changed when I was posted from number 56 Squadron to number 29 Squadron. I consequently returned from Cyprus to Wattisham, to help with the establishment of the new Squadron. By that stage, the delightful cat-suit babe and I were getting on well, and I therefore had another good reason to remain at Wattisham.

In June 1967, as promised, my Instrument Rating Examiner's Course began at Coltishall. I thus became familiar with the road between Wattisham and Coltishall.

On one of these journeys, I was returning to Coltishall on a Sunday evening, when disaster suddenly stared at me. An old van pulled out from a petrol station to cross the road, just as I was approaching. The driver saw me late, at which point he froze, blocking the carriageway.

All four wheels of my Jaguar XK140 were locked, but it was clear that a collision was inevitable. It felt somehow surreal, knowing that the crunch had to come. I had an instant in which to decide what part of the van to hit. I chose the rear, hoping it would cause least injury to the van's passengers.

It was a good decision; the van was full of children.

The Jaguar swept the rear of the van unceremoniously aside, and ended up pointing directly at one of the petrol pumps, still travelling quite fast. All four wheels remained locked.

The nose of the Jaguar eventually ended up about three millimetres from a petrol pump.

I gave a sign of thanks to above, and leapt out of the Jaguar to help with the children, who were all hysterical by then. A petrol attendant rushed out and squirted the contents of a fire extinguisher over the Jaguar's engine.

"Call the Police," I yelled.

The Police, when they eventually arrived, turned out to be a local Bobby on his bicycle.

Of unflappable Suffolk stock, the policeman succeeded in calming everyone down. Eventually, statements having been taken, the Jaguar was abandoned, and someone gave me a lift to Coltishall.

However, a problem remained. I was Jaguarless, and the question of wooing a new girl-friend played on my mind.

"How am I going to see the cat-suit babe?" I asked.

"No bother," said Mike, a fellow Lightning pilot. "You can borrow my Lambretta!"

The two-wheeler lacked the panache of the Jaguar. Nevertheless, it had to be admitted that it was a wonderful feeling Lambretting through the Suffolk country-side at the height of Summer. The narrow lanes, full of character, divided the acres of wheat fields. The sweet country smells were sensational. There was a feeling of warmth and relaxation. Nirvana was at the end of the journey.

Members of No 29(F) Squadron, 1967.

❖ ❖ ❖

"Richard…is it you?…" the scene was a busy London street. The date was the 1990s, around thirty years in the future. In the bustle of Knightsbridge, fatefully I had bumped into a charming, successful couple. They looked at me. I looked at them. The look of doubt soon materialised into recognition. She kissed my cheek. It was the cat-suit babe, and her husband, Mike, the Lambretta lad.

In the summer of 1967, however, my crystal ball did not stretch that far into the future. At the time, I did not realise that my relationship with the cat-suit babe was not destined to be long-term. The parental pressures and self-doubts lay a few months ahead yet.

❖ ❖ ❖

'*Yachting and Craft Monthly…July 1967.*' It was one of the obscure magazines left lying around in the Mess at Wattisham. By chance, I picked it up one day. The advertisement for a small Cabin Cruiser looked intriguing. I telephoned the number given. A gruff but friendly voice gave more details. The speed-boat had a powerful engine, enough to water-ski, it also had a small cabin, and came complete with towing trailer and accessories.

A trip to the Malvern area soon sealed the bargain, and the newly-restored Jaguar towed the Cabin Cruiser back to Wattisham. The new possession was installed at Bawdsey Ferry, near Felixstowe, and my long-suffering girlfriend would take the wheel while I mastered the art of water-skiing.

"You pilots!" she said. "You're speed mad, and always on the go. It's crazy!"

"Sorry," I said. "For me, this is relaxation away from the flying. And you're such a cool driver." Her beautiful eyes narrowed as she looked at me. She laughed.

After a while, I mentioned something else: "We've just had the most tedious Exercise," I moaned.

Then I told her about the rebellious radio-call incident.

It was the middle of the Exercise, and a row of Lightnings stood on the dispersal area. Each aircraft was manned, and the pilots were waiting for instructions to scramble. Generally impatient by nature, the average Lightning pilot soon became bored. Action was preferred.

"Who dat?" The call was from a Lightning pilot, but we were not sure who it was.

"Who dat?" The call was the same, but the voice was different.

In no time, a collection of "Who dats?" was bouncing around the air-waves.

"Wattisham aircraft, this is the Wing Commander Operations. Unauthorised calls are to cease immediately. What the blazes do you think you're up to anyway?"

There was a respectful pause, then: "Who dat?"

"*Wattisham aircraft this is the Wing Commander Operations…*oh, to hell with it. All aircraft *…scramble, scramble, scramble.*"

As the Lightnings filed out in rapid, but orderly succession, there were a few grins and thumbs-up signs between the pilots at the canny way in which we had manipulated action.

❖ ❖ ❖

The formation of the new Squadron, number 29, progressed quietly and successfully during the following months. Keen to expunge the Batman and Robin image, an efficient team laboured to develop new ideas. As the new Instrument Rating Examiner, my speciality was to ensure high standards in the pilot's instrument flying techniques. It was an important aspect in the Lightning world, where the type of work relied on good instrument flying.

However, my time with the new Squadron was not fated to last long. In the Spring of 1968, rumour abounded that an Instrument Rating Examiner was needed on one of the Lightning Squadrons in Germany.

When the posting notice came through, I received it with rather mixed feelings.

The girl-friend problems had become difficult by then, but even so I was dreading having to part. Various ideas were discussed, but we both knew in our hearts that they would not work.

The final meal, at her home with all the family, was sad and awkward. The beautiful eyes drooped. The eyes looked damp.

As one who hated prolonged farewells, I felt inept and embarrassed. Finally, I walked out of the household. It was dark outside, and I felt the chill of the night air as I strode to the Jaguar XK140 parked in the Ipswich street.

When I was walking to the car, I suddenly heard a muffled wail from behind me. The sorrowful tone drifted through the night air. It was distinctly heard, and caught a note in my mind, audible to this day.

As I drove away through the Town's suburbia, I reflected on the happy times which had dominated my life so recently. Suddenly, it had all ended. Even the Jaguar, that wheeled witness, gave the illusion of shaking its head sadly.

CHAPTER 12

Welcome to Gütersloh

I stood up as the tall Wing Commander walked over to where I was reading a newspaper. It was March 1968, and I was in the Officers' Mess at Gütersloh, a large airbase in the northern plains of Germany. The dark-haired Wing Commander looked at me with an air of amusement. With his head a little to one side, he asked my name.

Then Wing Commander Laurie Jones said: "Welcome to Gütersloh!" The Commanding Officer of Number 19 Squadron shook hands with me firmly. I felt impressed; it was the first time (and the last) that a Commanding Officer had taken the trouble to root around to find me, the new Officer on his Unit.

We chatted for a while, then he said: "Once you've settled in, there are a number of Instrument Ratings for you to do. We're a bit behind because of sickness and other problems."

After introductory briefings and presentations about the Squadron's role in Germany, I had some familiarisation flights. Then I was asked if I was ready to start dealing with the Instrument Rating backlog. The Instrument Rating was flown in two parts, using the twin-seat Lightning T4 Trainer. High level turns and manoeuvres on the first part of the Test were flown subsonically. For the second part, the Test included supersonic turns at Mach 1.3, a difficult exercise for the candidate to fly accurately. At the end of the sortie, the candidate had to fly an 'Instrument Landing System' or a 'Ground Controlled Approach'. A successful annual Test was needed for the award of a 'Green Card' certificate, which qualified a candidate as competent.

"This guy's a right chatter-box," a candidate on one of these Tests, Les, said at the end of his Instrument Rating flight. He referred to the Air Traffic Control Officer.

"Let's speed him up a bit then," I said.

"How do you mean?"

"Well, we've got enough fuel for another Ground Controlled Approach."

"So?"

"When you've settled on finals, and he's started the talk-down, just raise flaps and undercarriage, apply a quick burst of reheat, and we'll be at 400 knots in no time."

"Right on!" said Les. From the East End of London, Les was known to be a rather eccentric character. Amongst other things, he enjoyed telling us about his private Fire Engine kept in a London lock-up.

We initially followed the Controller's instructions at normal speed, and were soon settled on the final approach.

"Check wheels down and locked," said the Controller. "Turn right 5 degrees. Commence rate of descent for a three degree glide-path."

Les looked across at me. I nodded. The wheels and flaps were raised, and he applied full reheat for a moment, before throttling back to maintain the speed at 400 knots.

"Turn right a further 5 degrees," continued the Controller. "You are on the glide-path. That's a good heading." The Controller's pace was getting faster. "You are going slightly above the glide-path...turn left two degrees...back on the ..." The Controller was even faster: "*Turn left another two ...increase your...turn...*" The Controller had finally run out of steam. "Lightning 20...what's going on...?"

"Sorry about that," said Les. "A slight slip of the throttle. We'd better join the circuit visually. Bye."

After we had landed and shut-down the Lightning T4, I saw Laurie Jones coming towards us.

"Had a complaint from Air Traffic Control," he said.

"Not about us, surely?" we asked ingenuously.

"Surprising isn't it?" he replied sarcastically.

"Can't understand it," we said.

"Tell you what," he said. "You two can do Battle Flight tomorrow." Les groaned.

The Author's Lightning F2 over Gutersloh, 1968.

The 'Battle Flight' hangar at Gütersloh housed two fully armed Lightnings. It was a 'Quick Reaction Alert Force'. As a lot of the time was spent sitting around, it was not a popular assignment.

If the 'Battle Flight' alert siren sounded, we had to be airborne within five minutes. Gütersloh was quite close to the border between East and West Germany. A 'buffer zone' was heavily marked on all the aviation maps, and no-one was permitted to cross the border itself. Occasionally, light aircraft strayed over the border, and we were 'scrambled' to act as escort. On other occasions, we would set up 'Combat Air Patrols' paralleling the border. Usually high-level contrails could be seen across the border mirroring our flight path; Soviet Mig fighters of Eastern Germany made their presence felt.

"Better not upset Air Traffic Control today," Les said when we duly sat in the Battle Flight hangar performing our penance. "I can't understand why they're such a touchy lot anyway!"

Laurie Jones did not take long to calm down after the Air Traffic Control saga, and just a few weeks later he came up to me. "I've got a special assignment for you," he said.

I was to join a NATO Tactical Evaluation Team to evaluate a Royal Netherlands Air Force base in the northern part of Holland.

It was arranged for all the members of the Team to meet for a briefing at the Headquarters base at Rheindahlen. After the briefing, the Team were driven up to Holland. When we entered the airbase, we did so as discreetly as possible in order to maintain an element of surprise.

A local military car drove me to the 'Quick Reaction Alert' hangar, where I was to act as an observer. As I was stepping out of the car, the alert siren sounded.

I hurried in to the Operations Room, and showed my identity pass. The two pilots were running to the cockpits of their single-seat F104 aircraft, ominously dubbed the 'widow-maker' by many in NATO.

As the pilots waited, I watched them. Oxygen masks were clamped to their faces. They were checking round their cockpits. The F104 was pencil-shaped, with short stubby wings, and just one engine. Powerful electricity generators droned in the back-ground.

It was night time, and the atmosphere seemed sombre, almost ghostly. After a while, the pilots started to look around as they continued to wait in their cockpits. One of them nodded as he caught my eye.

Eventually, a loudspeaker crackled into life: *"QRA aircraft, this is Fighter Control. For Tactical Evaluation exercise, scramble, scramble, scramble. Acknowledge."*

"QRA 1"

"QRA 2"

The pilots went into a well-rehearsed routine. There was a loud noise as both F104s started their engines. I wrote down times on my observer's note-pad.

The lead aircraft nosed out of its specially designed hangar, closely followed by number two. The glow of reheats from the rear of the two F104s soon disappeared as they took-off into the night.

It was time then for me to leave the 'QRA' hangar, in order to observe at the Squadron Operations Room. There, the activity was hectic: crew rotas were organised; briefings were arranged; programmes were worked out.

Soon we gathered for a meteorological briefing. The weather man pointed to his charts and tables as he described the area's synoptic situation. A clear night was anticipated at low levels, but extensive layers of medium and upper level cloud were forecast.

Suddenly, someone tapped me on the shoulder. "Telephone," he said.

I walked to the Operations Room, and picked up the telephone receiver. I was quite surprised to get a call. "Hello. Who is it?" I asked.

"Oh, hello! It's the Team Leader here. Can you confirm that both QRA aircraft scrambled, please?"

I asked him to wait one moment while I checked the details on my note-pad. There had been a sign of tension in the Team leader's voice. I felt a little anxious as I gave him the times.

"Is there a problem?" I then asked. There was a pause.

Eventually, the Team leader said: "Don't release this yet, but I'm afraid that only one of them has returned."

As requested, I said nothing, but perhaps my expression gave things away. As the crews filtered through to the Operations Room, an uncomfortable feeling had started to grow. An uncharacteristic atmosphere of apprehension had crept in.

At length, the Dutch Squadron Commander asked us to re-assemble in the Briefing Room. He looked weary and haggard. He broke the news calmly. Wreckage had been found already. The young pilot had not ejected. The Tactical Evaluation had been cancelled with immediate effect.

The Team Leader then requested his Team to rendezvous in the Officers' Mess. He attempted to placate the feeling that somehow the accident had been the Team's fault. We were told that the cause of the accident was being investigated. Team members, however, spoke amongst themselves about their suspicion that the pilot might have become dis-orientated while flying in cloud.

Eventually, the Team were taken back to Rheindahlen. From there the Team split-up, and I drove up to my home base.

When I reached Gütersloh, I was keen to fly as soon as possible. "Just get me airborne..." I said to the Duty Operations Officer. "It'll take my mind off things." We both knew that it was the quickest way to re-build confidence.

I was glad to climb into the familiar Lightning cockpit. The instrument lay-out, the rows of dials and switches, the twin throttles, even the hard ejection seat were now well-known and friendly territory. With the rumours about the F104 pilot's possible dis-orientation, I felt more than ever duty-bound to strive for the highest standards of instrument flying on my own Unit.

Some time after my return to Gütersloh following the F104 tragedy, I was approached by Laurie Jones. "As you fly the T4 trainer a lot," he said, "Perhaps you're a good person to fly our visitor on a familiarisation flight."

Two aircraft from a Danish Fighter Squadron were visiting us. One of the pilots, Lieutenant Christenson, had asked for a flight in a Lightning.

He took a keen interest in the pre-start checks, and start-up procedures. As we taxied out, I fielded his questions. "I can't let you fly the take-off," I said. "But once we're at a safe height, you can take the controls."

I applied full cold power for the take-off run, and waited a few moments before rocking the throttles outboard to the reheat position. There was the usual momentary pause, then *wham!* as the reheats kicked-in. It had become second nature to me by then. I rotated the aircraft, and kept the reheats engaged to demonstrate the rapid rate of climb.

Suddenly, as we climbed away from the airfield, there was a *"clang... clang... clang"* through my head-set, and I was confronted with a flashing red light in the cockpit saying 'FIRE 2'.

Immediately, I dis-engaged both reheats, and throttled back number two engine to the 'idle' position. At the same time I called: *"Mayday... Mayday... Mayday... Dolphin 01... Fire in the air... Standby."*

The fire warning persisted, and a check in my rear-view mirror revealed signs of smoke. I therefore shut-down number two engine, and pressed the fire extinguisher.

After some moments, the fire warning light began to grow dim. I checked in the mirror again, and noted that there was less evidence of the smoke.

"OK," I said to my passenger, "I'll have you down in no time. But just in case, remember the ejection seat drills we discussed." I was already in a descending turn back towards the airfield.

"Gütersloh Tower, Dolphin 01, the fire appears to be out. Downgrading to PAN. Request priority landing."

As the runway edge approached, I landed Lightning XM 991 firmly, and we felt the tail parachute billow out behind us. At the end of the landing run, I taxied the aircraft clear of the main runway and closed down the good engine as fire crews rushed towards us.

"Sorry about that," I said to Lieutenant Christenson.

"It was a pretty quick flight," he replied.

We had logged just 5 minutes from take-off to landing.

An Operations Truck then drove us away, and I was taken to speak with the engineers. After that, Lieutenant Christenson and I walked to the Operations Room. "What was that all about?" asked Bob, the

Duty Operations Officer, as we entered the room. I told him about the drama.

"Fire in the air," whistled Bob. "And with a passenger too. Bad luck." He then volunteered to help me with the form-filling and other paper-work.

"By the way," said Bob eventually, as we filled in the reports, "who was that blond girl I saw you with last week-end?"

"Who me?"

"Yes, you."

"I had a letter from General Harlinghausen, who lives in the town," I said. "He's a retired Officer from the Luftwaffe, and had heard about my posting here. He said that he and my father had been adversaries during the War, but to show there were no hard feelings, he invited me to his home to drink some coffee, and to meet his family."

"And the girl?" asked Bob.

"His daughter."

"Wow."

With long blond hair, a slim figure, and bright blue eyes, Gisela was an archetypal German beauty. She was a medical student at Gottingen University, close to the border with East Germany. Having lived in the USA for a year, her English was fluent. We seemed to make good companions, and it was perhaps inevitable that I should have found myself driving over to Gottingen as we got to know one another.

We dined in atmospheric 'Gastatte' with candle-lit tables, and zither music in the back-ground. The old-world charm of Gottingen itself, with cobbled streets and beamed buildings, made a romantic setting. Outside the town, we went on long walks through forests with colourful trees, tall and disciplined-looking, typical of wooded areas throughout Lower Saxony.

Like many young Germans at that time, Gisela felt a confusion about the relationship of her country with other nations. World War 2 had not been the fault of their generation, they had not even been born, but still they felt a sense of guilt. In neighbouring countries, some of the older folk harboured a hatred for Germans. My serious-minded friend loved to spend hours debating the topic.

Once, when it was time for me to leave, I was driving Gisela back to her flat, when a bright flash from an over-head camera was followed by a traffic policeman waving us down. My wallet was suitably emptied to pay an on-the-spot fine, and she then looked across at me.

"I suppose Germany has just made one more enemy," she said.

"Afraid so," I replied.

"Well, one thing will cheer you up," she said. She glanced at me, and smiled. "At least you can look forward to flying those Lightnings next week!"

Members of 19 Squadron with a Lightning F2A; Gutersloh, Germany, 1965.

CHAPTER 13

Missile Mayhem

"Make your heading 175 degrees."
We had been briefed on the need to fly with great accuracy for the 'Missile Practice Camp' profiles.

"Turn right 5, onto a heading of 180 degrees," called the Controller. I nudged the Lightning's control column slightly to the right, and concentrated on holding the aircraft steady, so that the compass heading was exact. Flying over Cardigan Bay, the Welsh mountains were clearly visible way over on my left side. It was a clear day, and the sun reflected brightly from the sea. The conditions were good for the practice missile firing about to take place. The missiles themselves were valued at around £1 million a-piece; the opportunity for a 'live' firing was therefore given to us only rarely.

"Turn further right onto 185 degrees." I moved the control column again.

Our missiles, code-named 'Firestreak', had infra-red sensors fitted in their nose. When within range of a target, these missiles would 'lock on' to the infra-red source, usually the jet engine efflux from an enemy target. That day, however, in the firing range, a Jindivik radio-controlled aircraft was towing a special infra-red light. The intention was for the Firestreak to 'lock on' to the towed light, and not the Jindivik itself, which was valued at considerably more than the missiles.

"Hold your heading on 185 degrees." The Controller's voice was calm and matter-of-fact.

"Dolphin 19, your target is now to your left, range 15 miles, heading due west. Acknowledge."

"Dolphin 19 acknowledged." This call made me feel a little nervous. The Controller was saying, in effect, "This is it, guys. Do NOT screw up!"

I looked over my shoulder, to the right. I could see the 'chase' aircraft, another Lightning, in loose echelon formation. His job was to capture the missile firing on film. Beyond him was a third aircraft, as safety back-up. He also carried camera film.

"Your target is now range 9 miles. Turn further right onto heading 190 degrees."

I checked and double-checked the switch positions in my Lightning cockpit. For the missile firing, there was a trigger under my fore-finger. At that moment the safety-catch was firmly applied. Peering into the Lightning's radar screen, I could distinguish the Jindivik as a 'blip.'

"Your target is now range 5 miles, stand by for right turn onto final heading."

I could see the target visually then, just crossing the nose of my Lightning. A small Jindivik aircraft, with short wings, was towing a long cable at the end of which was a bright source of light.

"Target now range 4 miles. Turn right onto 270 degrees."

I followed the controller's instructions. As I rolled out onto a westerly heading, I was aware that 'nerves' were playing a part; my flying was not as smooth as it should have been. "Cool it!" I told myself.

"Target range now 3.5 miles."

The Jindivik, and its towed target, were clearly visible ahead, and slightly below me.

"Target range now 3 miles. Standby Dolphin 19...*Standby!*"

"What is this Welsh twit on about?" I thought. "I've been standing by for the last 10 minutes."

"Target range 2.75 miles."

"Standby...*Standby*..."

There was a pause, then: "Check safety catch."

I released the catch.

"Target range 2.6 miles. *You are clear to fire!*"

I made a final check of cockpit switches, but did not squeeze the trigger immediately. There was a 'window of opportunity', and I had a few seconds in which to make sure that all parameters were correct. As well as ensuring that my cockpit switchery was faultless, I wanted to guarantee the correct positioning of the other two aircraft. I glanced in the rear-view mirrors, and noted that both of the 'chase' aircraft remained in place. Ahead, I could see the Jindivik still, and behind it the bright light of the towed target. I checked the radar picture once more, and re-confirmed my height on the altimeter. Everything seemed in order. Just a small movement of one of my fingers was the final requirement in a long chain of events; momentarily, the thought struck me as bizarre.

Then I squeezed the trigger.

The intercept profile used on that day had been practised many times; it had become familiar. Normally, however, when the trigger was pressed on a training exercise, with just dummy missiles fitted, there was no end result. I therefore had a feeling of surprise, almost shock, when there was a loud *whooooosh*, and the whole aircraft rocked.

I blinked. For a fraction of a second, I thought: "What have I done?"

Just ahead of me, I could see the slender shape of the white Firestreak missile accelerating away on its one deadly flight. Smoke was streaming from the rear of the missile. Another quick glance in my rear-view mirrors confirmed the presence of the 'chase' aircraft. The Firestreak's rate of acceleration increased all the time, but I could still spot the missile visually as it pursued its lethal course.

Suddenly there was a flash.

"Bingo," called the chase aircraft. My eyes strained ahead as I tried to find the Jindivik. However, there was nothing to be seen.

"Dolphin 19 from Range Controller. Confirm firing complete." The Controller's voice interrupted my thought process.

"Affirmative, Dolphin 19," I replied after a pause, "but there's no sign of the Jindivik."

"Oh dear!" said someone.

"Dolphin 19 commencing search." I started to orbit around a spot in the sea where smoke and steam could be seen. Bits of wreckage were floating. In spite of my oxygen mask, and having 100% oxygen selected, I noticed unpleasant, distinctive smells; a mixture of burnt rubber, aviation fuel, and hot metal.

Eventually, I called the Range Controller with a terse message: "Dolphin 19. Jindivik has been destroyed. Returning to base."

I applied power to the engines, and the other two Lightnings followed as we flew back to our base (Valley in Anglesey) at a speed of around 400 knots. We overflew Valley, and turned downwind into the 'circuit' procedure.

After landing and shutting down the Lightnings, we walked to the 'line hut', to sign the engineering log-book.

"How did it go, Sir?" asked the engineering 'Chiefie'.

"Afraid I shot down the Jindi," I told the Chief, as I signed and dated the log-book.

"Oh dear," he said. "Never mind, Sir, not your fault. Not your fault. It's a big Company anyway!"

We waited eagerly for the 'cine' film to be developed. It did not take long, and we inspected the results minutely. The Firestreak missile could be seen leaving the Lightning, dropping slightly, and then accelerating away. We made out smoke trailing behind the missile, and then saw the missile stubbornly by-passing the towed target, and heading directly for the Jindivik. We slowed down the 'cine' film to watch, frame by frame, the fatal last moments of both missile and Jindivik. The Firestreak broke up in flight at precisely the right moment to ensure that missile and target exploded together.

The Range Controller was philosophical when he rang. "We'll just have to try a bit harder tomorrow," he said.

On our return to Gütersloh at the end of the Missile Practice Camp, the comments were probably inevitable: "That'll cost you," "Mind your Mess Bill."

But Laurie Jones had other ideas.

"How'd you like to lead a four-ship formation?" he asked me. It would be part of a local 'open day' which was planned that year, the last in the decade of the 'swinging sixties'. My job would be to lead four Lightnings in a short flying display over the airfield.

"That's fantastic. Thank you," I replied.

"Every dog has its day!" said Laurie Jones.

The first practice was at Gütersloh, on a fine morning in that early part of summer.

"Dolphin check in!" I called on the radio.

"Dolphin 2."

"Dolphin 3."

"Dolphin 4."

I called Air Traffic Control for clearance to taxi. We were cleared then to line-up on the runway for take-off in two pairs. "We'll avoid all four aircraft taking-off in close formation," I had briefed, "in case of reheat problems with one of the front aircraft."

As I nodded firmly, it was the signal to my number 2 to release brakes, and follow me at the start of the formation take-off. A further nod meant 'engage reheats'. Soon we were flying in a gentle left-hand curve, with my number 2 tucked tightly into the echelon starboard position.

We throttled back the engines to a relatively low power setting. That gave the other pair room for manoeuvre as they closed up to their echelon port, and line astern, positions respectively.

With the four Lightnings in place, I cleared the circuit area, and climbed to around 5,000 feet. We flew towards Bielefeld, to the north-east of the Airfield, and I chose a clear area just before the high ground of the Teutoburger Wald range. At first, I led the formation into wing-overs, the well-proven warm-up manoeuvre.

As the formation coped well, I made the wing-overs steeper. My aim was to fly smoothly, using as little throttle movement as possible. That allowed the other aircraft margin to make constant fine adjustments to their position. Their eyes concentrated entirely on looking at the lead aircraft. It was therefore my job to attend to navigation and airspace clearance.

"Standby for barrel roll," I called.

I eased my Lightning into a shallow right-hand dive. Then I applied 'g' to pull the nose of the aircraft above the horizon, and reversed the direction of turn. That turn was continued through 360 degrees, and the 'g' force was kept constant.

"Great!"' I thought. "We'll out-do the Red Arrows yet."

Soon, we had put together our short display, and I called Air Traffic Control for clearance to return to the airfield.

I felt a surge of excitement as we ran through the display. "Surely the peak of any pilot's career," I thought to myself. When the routine was complete, Air Traffic Control gave permission for a final fly-past, after which we asked to 'break into the circuit'. As we taxied our aircraft to the dispersal area, I saw Laurie Jones' ubiquitous figure waiting to greet us.

"That was pretty good," he said. I noticed a long list of notes he had made whilst watching the display. The de-briefing session took quite a long time. "Overall that was OK," he said at the end of the de-briefing. "Don't get complacent, but well done everybody."

When the de-briefing session was over, the Commanding Officer took me to one side. "I've got some news for you," said Laurie Jones. I looked at him. "I heard today that you'll have a few more months here to finish your tour, and then you can expect a posting notice next Spring. It looks like they've got a ground tour lined up for you."

My heart sank.

Lightnings parked, Gutersloh, 1968.

CHAPTER 14

Auf Wiedersehen Lightnings

As I checked the flying programme one day, a voice from behind said: "A little jaunt for you tomorrow." It was the Duty Operations Officer. I looked at him doubtfully. "Take XN 727 to Warton, and bring back XN 735."

At Warton in Lancashire, near Blackpool, was a factory owned by the British Aerospace Company. The scheme was to fly a Lightning Mark 2 from Gütersloh to Warton, where it was modified to Mark 2A standard. The improvements included a larger fuel ventral tank, fitted under the belly of the aircraft. The delivery of the Mark 2, with its smaller ventral tank, consequently was made in two stages; the return flight in the Mark 2A Lightning was made in one step.

"The outbound Flight Plan stipulates a refuel at Wattisham, and then you fly direct to Warton" I was told.

Eventually, after touching down at Warton in my Lightning Mark 2, I climbed out of the aircraft. "Not often I get a welcome party," I said to the member of British Aerospace management.

There seemed to be no hurry, and I was ushered into a plush cocktail bar full of people.

It had not occurred to me to take a change of clothing, and I felt rather out of place standing amongst the smartly tailored suits. I was in my basic Service Flying Kit '*Aircrew for the use of...*'

The Directors present did not appear fazed by my state of dress, in fact they seemed to enjoy the novelty factor. Discreetly, they almost made me a centre of attention. I was asked penetrating questions: "As an operational pilot, what is your opinion of the radar improvements… What about the new Artificial Horizon presentation… Would it really help if we modified the reheat latch?"

"Why that?" I asked.

"You haven't heard?" they replied.

"Heard what?"

"About the Engineering Officer at the Maintenance Unit?"

"No."

"A test of a Lightning's braking system nearly ended in disaster."

"Tell me more," I said.

"The Unit Test Pilot wasn't available for some reason, so the Wing Commander in charge decided to do the brake check himself on the stripped-out Lightning. After a short test, the brakes seemed fine, so he elected for a longer run."

The Wing Commander pushed the twin throttles fully forward. Perhaps he got carried away. The noise would have been intense, especially as he wore no headset.

"He got the throttles stuck in reheat," said my hosts. "And the aircraft got airborne. He was just sitting on an orange box, in shirt-sleeves, with no oxygen, head protection, or anything."

"Scary," someone said.

"He wasn't even strapped in. There was no canopy, and he had no form of communication."

"Amazing," commented a listener.

"The Wing Commander figured out eventually how to release the reheats, but by then he was at 30,000 feet. The temperature was around minus forty, and he was becoming anoxic. Using techniques from his Cessna private flying, he managed to point the Lightning back towards the airfield, and he started a descent. During the descent, he circled the airfield to keep it in view. At length, he decided to attempt a landing. He was high over the runway edge, but forced the machine down for a crash landing. Ironically, he proved that the brakes were effective: he achieved the shortest Lightning landing on record, bursting both tyres. The fire crews had to force his fingers off the brake lever, one by one, before lifting him out of the cockpit."

"Poor chap," someone remarked.

"Unbelievable!" said another person.

As we walked into an executive dining room, the story continued to dominate our conversation. "Even James Bond would be stumped by that one," and other such comments.

It was heard (some months later) that the Wing Commander in question had eventually suffered a nervous breakdown and, regretfully, had to be discharged from the Service.

When lunch was over, after polite farewells to the executives, I was driven away. As we drove past the British Aerospace Company hangars, I spotted my return transport to Gütersloh: a gleaming Lightning Mark 2A, number XN 735, stood in solitary glory.

I had a feeling of excitement, a bit like the collection of a new car. I was also aware of something else: crowds of British Aerospace workers were streaming out of a hangar. They lined up to watch. At first I did not appreciate what was happening, but then the manager said: "They've spent months working on this aircraft. It's their pride and joy!"

As a matter of routine, I walked around the shiny Lightning, performing the external checks. I need not have bothered. Everything looked polished; the tyres were new; the canopy gleamed; even the ladder taking me into the cockpit was smart and unbattered. By now, the onlookers appeared to be in their thousands.

Inside the cockpit, everything was pristine. The paintwork was professional, the instruments were clean. As I went through the familiar start routine, every so often I glanced at the assembled crowds. It was an unusual situation. Normally the Lightnings were operated discreetly, well away from public gaze. As I taxied out, I raised my hand in greeting. I felt like a film star as a sea of hands waved back.

They still watched as I lined up for the take-off run. Just before the take-off, Air Traffic Control asked me to perform a fly-past before returning to Germany.

I carried out a slow fly-by over the watching crowd. As I flew overhead, I waggled the wings, and straightened the Lightning on an easterly heading. I engaged the reheats at the same time as pulling the nose upwards to an angle of about 45 degrees.

Within around two minutes I was levelling at 35,000 feet. Those watching at Warton would have seen white contrails forming behind the Lightning. Meanwhile, the Air Traffic Controller at Warton thanked me for the fly-by, and gave a radio frequency for the Radar Controller. I then called the Radar Controller at the London Air Traffic Control Centre.

"Radar, this is Dolphin 11, climbing out of Warton en-route to Gütersloh."

"Roger, Dolphin 11, good afternoon. Call me passing 5,000 feet."

"I am past 5,000 feet," I replied.

"Roger, call me passing 10,000 feet."

"I am past 10,000 feet." There was a slight pause.

"Roger. In that case call me passing 15,000 feet."

"I am past 15,000 feet." There was another pause.

"Roger, Dolphin 11. Can you confirm your height and position." I gave him the requested information.

"That's understood, Dolphin 11. Can you confirm your aircraft type, please." I told him.

There was a further hesitation from the Controller.

"Roger Dolphin 11. You might have mentioned it earlier."

"I couldn't get a word in edgeways!"

The flight back to Gütersloh took one hour and twenty five minutes. The flight proved to be one of my last in the Lightning world.

"Not much time left in Germany," Gisela said to me the following weekend. "Will you miss us?"

"You're a funny lot, but I'm afraid I will," I replied.

"Think of all those lovely British girls," said Gisela.

"I'll think of this lovely German one."

"You'll miss those flying machines more."

"It'll seem strange having my last flight," I reflected.

It was a day I dreaded.

On April 30th 1970, I flew Lightning Mk 2A, number XN 778, hull letter 'H'. My total flying time on Lightnings had come to just over 1,071 hours.

Gisela and I bade *auf Wiedersehen*, and made arrangements to meet in the United Kingdom. Number 19 Squadron held a traditional dining-out to say farewell.

Eventually, I drove to Headquarters 11 Group, at Bentley Priory, near Stanmore, in Middlesex; Laurie Jones had been correct about the ground tour.

The Headquarters was where my father had been Commander-in-Chief, when it had been Fighter Command. It was thus familiar territory in a way. Reduced in status, the Headquarters seemed a shadow of its former self.

In the Battle of Britain, Winston Churchill had visited the Operations Room, deep underground, and had watched as the WAAF operators moved plaques representing the disposition of Royal Air Force and Luftwaffe Squadrons. Loudspeakers relayed the radio calls made by the pilots. The calls, clipped and brief initially, became vociferous when the pilots started to engage enemy forces. Eventually, the language, and the shouts of the pilots fighting for their lives, became unbearable. The loudspeakers were then turned off.

My office had a label on the door which said 'Operations (Home)'. I had a desk, and a filing cabinet. It took a long time for me to adjust to the mentality of '4 square walls'. It seemed so unnatural, having been used to the freedom and excitement of the skies. The culture shock was depressing.

Next door, my superior officer had an office similar to mine. His superior had a slightly larger office.

My superior officer was also an ex-Lightning pilot, and had a highly competitive nature. He had the habit of rolling his eyeballs upwards, and fluttering his eyelids, if he cracked a joke, or said anything controversial. He wandered into my office at random, hoping to find me bored, or unoccupied, or both. He lapped up the paper-work and officialdom.

My lowlier bureaucratic status, not to mention ability, was highlighted occasionally. The key for coping, I discovered, was to have an effective system to look occupied and interested at a moment's notice. In my filing cabinet, therefore, I arranged various files and documents to be available in double quick time. I reckoned to have about four seconds warning between hearing the familiar footsteps, and the entry of 'himself' through the door.

At weekends, I usually drove to see my parents at 'Little Wynters', a house they had built in the early 1950s, near my Grandfather's house.

Gisela visited, as planned, and she tried to cheer me up. However, her visit somehow underscored cultural differences. When she left to return to Germany, there was not unbearable sadness between us.

It was at about this time that the wheel of romantic fortune came up with another unpredictable twist.

Some friends who lived near Little Wynters asked if I would play at a local tennis party; someone had dropped out. She mentioned, in passing, that my partner would be an attractive girl who worked in London.

In July 1970, one Saturday morning, I therefore set off to join the tennis party. Unfortunately, I had difficulty in finding the right place, but eventually drove up to a large country house to find that the tennis had started already.

As I walked through an orchard area, I caught a glimpse of a seriously pretty girl practising her tennis serve. I went over to the hostess, and offered apologies for being late. The hostess was delightful, and poised, but pointed out that I was needed on court straight away. My 'set' had started already.

It suddenly became apparent that the seriously pretty girl was to be my partner. "You're such a sucker when it comes to seriously pretty girls," I told myself. "Get a grip."

I walked onto the tennis court. There was bright sunshine. I introduced myself, and looked again at my partner. I could not believe how good-looking she was.

In between the tennis, and even on the court, I chatted easily with my partner. She seemed natural and unaffected.

When the tennis was over, my partner suggested that some of us should go over to her place that evening. Meanwhile, she drove herself off, and went back to her parents' home.

It would be many months before I found this out, but when she got home, at first she could not answer her parents' enquiries about how the day had gone.

"Are you OK?" her mother asked. There was still no answer.

Her father, always diplomatic, asked quietly: "Did you have a problem?"

She looked at her father. The atmosphere was rather tense; she had recently ended a relationship which had caused a lot of unhappiness.

"I think I have just met my future husband," Sue said to her parents eventually.

Gutersloh, Germany 1965 – inside the hanger.

CHAPTER 15

Just Married

My father and I were driving together to a family wedding one weekend, some months after the Tennis game, when he glanced at me and asked: "When are you getting engaged to Sue?" I felt a little touchy about his directness. Such a question was somehow untypical of my father.

"Well, actually, I asked her last night," I replied eventually. "But I haven't had an answer yet."

My engagement to Sue, however, was confirmed the following week.

It was an ecstatic period. "Not that we didn't have problems," Sue said to me one time. "Especially when that driver nearly finished you off."

One Monday morning, a military car had collected me, and two other Officers, to take us to Bovingdon airport, on the western outreaches of London. We planned to visit various Royal Air Force Stations within 11 Group.

Whilst trying to cross the dual carriageway into the airport, the driver made an error of judgement, and nosed into the heavy traffic.

"Watch out…" the yell from the passenger behind me came too late.

The car then stalled halfway across the road. There was an awful pause, followed by a loud crunch, at which point I lost consciousness.

My side of the car took the full force of the collision.

I remained unconscious while they dragged me out of the car, and took me to a Guardroom. I was getting glimpses of consciousness as I was laid down. "This one's quite bad." The policeman's words drifted through my semi-conscious mind as they man-handled me. I vaguely remembered feeling sick, and seeing my Number 1 uniform stained with blood. I recollected being lifted into an ambulance, still drifting in and out of consciousness.

Eventually I woke up in Hillingdon Hospital, ironically not far from Benbow Cottage. I had a headache, and the Doctors were worried about a burst spleen. Various people visited me in the Hospital, including my fiancée Sue, and both my parents.

Some weeks later, Sue and I went to look at the wreck of the car. "Lucky it wasn't a few inches further back," said the mechanic. "It would have been curtains. You had a guardian angel that day!"

Another problem arose during our engagement when I received a letter from Gisela. She was keen to try to patch something up. I agonised over the wording of a 'Dear Jane' letter; eventually I posted it, but received no reply.

In spite of these problems, the arrangements for our Wedding progressed happily. The Wedding Day was set for mid-summer – July 24th in the year of currency decimalisation, 1971.

The day was windy, but dry and bright. We were married in a small local Church, Magdalen Laver, an old-fashioned building constructed partly with flint. The Reverend Harry Roberts conducted the Service. "It's nice to marry a couple who are both Churchgoers," he said.

After the Wedding Service, we gathered outside the Church for photographs. The Bride's hand-made Wedding Dress, so carefully chosen, was flowing and beautiful. It billowed out in the gusty conditions. "Don't get me wrong," I whispered to my Bride as we smiled and waved to the Wedding guests, "but you're veil is definitely one up on a Lightning tail parachute."

"Thanks," she laughed. One of her sisters, Jenny, and her husband Colin, overheard the remark. They looked at each other in bemusement.

The reception was held in a marquee at my parents-in-law's home. Inside the marquee, protected from the wind, the Bride's beauty was paramount. She beamed as we mingled with the guests. The flower filled marquee appeared to nod in approval. The guests

Sue and I are married, 1971.

laughed and joked as they caught the flavour of deep delight felt by the Bride and Groom. Her parents' small dog wagged its tail contentedly.

My cousin, Hew, was best man. In twelve years time he would lead 3 Parachute Regiment in the Falklands War.

"I've known Richard a long time," Hew said in his speech, "but not Sue. It seems to me, however, that they are rather well suited."

The Bride and Groom were eventually driven to the Savoy Hotel, in London, for the start of their honeymoon. "Excuse me, Sir," said the Savoy receptionist. "Do you realise that you are leaving a trail of confetti behind you?" We walked over the thick-pile carpets, and were shown to our suite over-looking the River Thames. On the other side, we looked at the familiar Shell Oil Building, reminding me of the visit to see my father's office so many years earlier.

After a few nights in the Savoy Hotel, Sue and I drove to a large country Hotel at Portledge, on the north coast of Devon. After settling in to the Bridal Suite, we decided to try out the Hotel's swimming pool. The bathing-costumed newly-weds felt conspicuous as they walked through the hotel reception area, just as other residents had subsided into the copious Hotel arm-chairs for their afternoon tea.

"Allow me to assist you." One kindly gentleman had risen from his seat to help us fathom out how to open the large and quirky front door. The exercise was repeated in reverse around two minutes later after the Bride and Groom had felt the temperature of the ice-cold swimming pool (supposedly heated).

"Well that was a waste of time," said Sue. "We'd better try the tea, I suppose."

The wild coastline provided a romantic setting for the next few days as the young couple breathed-in the fresh Devonshire air. They walked arm-in-arm as they wandered bare-footed through the foam of the Atlantic breakers. Slowly they adjusted to their new situation.

❖ ❖ ❖

The first home we lived in was at Chipperfield in Hertfordshire. The house, called Greenbanks, was down a private lane, and had an attractive garden. Neighbours were friendly, and soon caught the drift of happiness felt by the newly married couple in their midst.

Every morning, I would walk down the private lane, and wait at the end for a military bus to take me to work.

As I walked into the Headquarters building, I was aware that the 'Robin' part of 'Batman and Robin' had been posted there too. He had a special car parking slot. However, as he lived in a nearby Married Quarter, he would cycle into work. 'Robin's' bicycle was usually placed ostentatiously in the middle of his car parking slot, so that no car would dare park there. As I walked into work one day, an Officer shook his head, and said: "It must be something to do with the size of mind!"

"Perhaps it is to do with the size of something else," someone replied.

Once in the Office, I dealt as quickly as possible with any files, so that I could then proceed with the more serious business of day-dreaming: *my magnificent, shiny, Lightning would be adorned with my amazing new Bride, and the two of us would taxi past throngs of admiring crowds, to... clump... clump... clump!* My four-second warning system was stretched to the limit sometimes.

As 'Himself' entered the office, if I looked a little guilty or flustered, it was detected immediately. I was supposed to be making 'Staff' comment on a bizarre piece of paper circulating the Headquarters. The Commander-in Chief of Strike Command, who worked in a building several miles from Bentley Priory, had scribbled notes one time, whilst on a flight home. The notes, literally on the back of an envelope, included a few thoughts about the 'Fighter' elements of his Command. These had been carefully photo-copied, and were sent round the various offices.

"It's a bit like the 'Thoughts of Chairman Mao'," one Officer remarked. Most members of the Headquarters staff found the platitudes quite irritating. Next door, though, managed to make something out of it. The paper he presented was impressive; it might even have ended up on the Commander-in-Chief's desk.

Occasionally, we travelled over to Headquarters Strike Command, to man the 'Air Defence Cell' during exercises. We spent hours sitting around doing very little. On a large map, placed in the centre of the Operations Room, a few WRAF members lethargically pushed plaques representing an aircraft, or a ship. It was never entirely clear what my duties involved.

"The main thing is to look keen and active," commented an old hand wryly. "You must try to know (or pretend to know) exactly what's going on."

The vitality, and urgency, which must have permeated the Operations Rooms during the Battle of Britain was sadly lacking.

When my superior Officer was posted, and the new incumbent arrived, it was quickly apparent that the new man was a different kettle of fish. In fact, he was of a remarkably similar frame of mind to my own.

"This place is a complete waste of space," he let slip under his breath one time. "It's so frustrating knowing that our skills as Lightning pilots are becoming more and more rusty as the months roll by." These skills had been so carefully, and expensively, honed.

Eventually, however, my time for posting began to draw near. I was summoned to the Air-Vice Marshal's office for my final interview.

"Ah, Richard," he said. "Sorry to be losing you. I'm sure you've found your time here most valuable."

As we finished chatting, he let slip a remark about my turning down the chance to be his *aide-de-camp*. The suggestion had been put to me whilst I was flying Lightnings in Germany. I might even have been quite tactless in my reaction to the offer. The whole incident had slipped my memory, but at that moment I had a sudden sinking feeling as it became apparent that offence had been taken. It made me feel quite gloomy, especially as I had liked the Air-Vice Marshal. I had found him to be a decent man; we had got on well.

However, I was determined to look on the bright side of life. I had been posted back to a flying job. I was destined to fly the F4 Phantom, based at Leuchars in Fife.

"The best thing of all," the Air-Vice Marshal had pointed out as he wished me good luck, "is the new bride you'll be taking with you to share the excitements ahead."

CHAPTER 16

A Pilot Again

"It'll take about a year," the Briefing Officer had told me.

"You'll start in Lincolnshire, take yourselves to Devonshire, before returning to Lincolnshire," he paused. "Then you and Sue can make your way to Bonnie Scotland."

Our first port of call, Manby, was near the Lincolnshire coast. We arrived there in November 1972. A remote, windy spot, it was the base for the School of Refresher Flying. The students 'refreshed' on the Jet Provost aircraft.

After some ground school training, my first flight felt strange after the two year absence. One side of my brain remembered the slickness of Lightning days, the other side was still stuck into the office mentality. Procedures which had been automatic when I was in current practise, were laboured and confused. Cockpit checks had to be learnt again, and even operating the radio caused problems.

The first few flights with my Instructor made me feel awkward. He reminded me of Cranwell days, with an immature approach; even a slight chip on the shoulder. "Here we go again," I thought. "What is it about these guys?"

By then, however, I was less reluctant to make my feelings known. An older and more mature Instructor consequently took over.

"I'm the solo aerobatics display pilot around here," the new Instructor said to me.

"Impressive," I replied.

"Perhaps," he said.

When practising flying on instruments, I was usually under a 'hood' to simulate flying in cloud. Part of the exercise involved recovering from 'unusual positions' (normally upside-down). In order to get the Jet Provost into an 'unusual position', the Instructor took control and flew manoeuvres. He then handed control back to the student to recover the aircraft to straight and level flight.

"If you don't mind, I'll just practice my aerobatic routine," said my new Instructor.

That took over 10 minutes. Having been away from flying for so long, I was still finding my 'sea-legs'. Rough flying made me feel airsick. My Instructor's flying, however, was highly polished.

"I didn't feel ill at all," I said. He beamed.

After the flight, I drove back to Sue. My journey took me through the unexciting countryside of Lincolnshire as I headed towards the coast.

"We could fix up accommodation at Mablethorpe," I had told her, some weeks ago.

"Mable-where?" she asked.

"Thorpe," I replied.

"Sounds fascinating," she said, as her eyes rolled upwards.

Mablethorpe was on the Lincolnshire coast. Once a holiday Mecca, the Town had slipped into the position of a sea-side non-entity. It was practically deserted in the Winter months. We lived in a terraced one-bedroom bungalow. The only form of heating was a paraffin stove. In the evenings, we sat close to the paraffin stove, huddled round, and listened to favourite radio programmes (we were denied the luxury of a TV). Condensation poured down the garish wallpaper, in places peeling from the walls.

There was a permanent gale blowing, and occasionally we tried to walk along the sea-front. Run-down, eerie, buildings lined the promenade; most of them seemed empty. A peculiar smell permeated the atmosphere: a mixture of salty air, frying food, and general decay. The few shops that were open sold cheap, downmarket items. The Town seemed frozen into a form of down-at-heel time-warp.

We had only one car, which was needed to take me to work.

"Sorry about the lonely days," I said to Sue.

"You didn't mention this part when we got married," she replied.

She would walk for hours along the fine sandy beach. The enormous beach appeared to stretch into infinity. Echoes of former glory filled the mind's eye as ghosts of children-past could be imagined running along the sand, screaming with delight. Desultory seagulls squawked above.

"I've got a solo flight tomorrow," I told Sue one time. "I'll try to 'buzz' you."

On more than one occasion I spotted a solitary figure wandering, and waving, as I swooped over-head. It made me feel quite guilty.

"Not much more of this," I promised Sue towards the end of my course. "We'll be off to glorious Devon soon. Then you'll see some life!"

"Farewell Mable-thingy," said Sue when we finally departed.

Our journey took us to Croyde, near Barnstaple. "This is more like it," said Sue as we drove through the Town.

The rocky coast, and rough breakers along the shore-line, complemented a warmer and more interesting character. There was a good atmosphere. We rented a small flat in a converted mews. The other flats in the block were occupied by fellow pilots; the owner's brother was in the Royal Air Force. The attractive local beaches reminded us of our honeymoon at Portledge. Sue had some companionship with other wives in the mews, and there was an active social life in the Officers' Mess.

"To get you back into 'Fighter Pilot' mental mode," the Flight Commander at Chivenor had said, "We'll do some formation exercises with four aircraft in low level 'battle' formation."

The Hunter aircraft were painted in camouflage colours. With their sleek shape at low level, the Hunters could slip around the countryside, using natural terrain to hide their position.

"The object is to see a potential enemy aircraft before he detects you," said the Instructor. "Two Hunter aircraft will fly line abreast, with a few hundred yards between them. Swept back from each of the lead aircraft, in loose 'echelon' position, will be the other two aircraft. We'll practice 90 degree turns initially, which will involve the two pairs crossing over each other."

Any other aircraft sighted, even airliners, had to be reported to the formation leader in the correct way, for example: 'Red 3, right, right, 2 o'clock high, range 8 miles, one aircraft crossing right to left.' All eyes would swing in that direction, and the leader would decide what action should be taken.

"We'll remind you about 'cine-weaving'," said the Flight Commander after the formation practices. "Two of you will climb to a height of 10,000 feet, and one aircraft will act as 'target'. The other will appear from behind, and aim to keep his gunsight on the 'target'. Be warned, it gets quite tricky when the 'target' starts to weave."

"The 'attacking' aircraft will use a manual twist grip to vary the size of his sight picture," continued the Flight Commander. "Depending on the range, the pilot will aim to put the wing-tips of the 'target' just touching the small diamonds on his gun-sight display. When within firing range of the guns, the 'attacking' pilot will squeeze his trigger which will start the 'cine' camera."

"When the sortie's over, the 'cine' film will be assessed by the staff."

The students spent several flying hours perfecting their skills at cine-weaving. By the Spring of 1973, however, the low-level flying and 'cine-weaving' at Chivenor had been completed. "I'm cine-weaved out," said a course colleague. "Time for a move."

As Sue and I packed up the car again, she said: "I'll miss the cream teas. By the way, how far will we be from Mablethorpe?"

"Miles away," I replied.

"Phew."

"Mind you, we could always visit for old times sake," I said.

"I'll look forward to the treat," she replied.

We were destined to spend four months at Coningsby, the base for the Operational Conversion Unit for the F4 Phantom. This aircraft had two engines with reheat, and carried eight missiles, quadruple the number carried by the Lightning. There were two members of aircrew. I had to fly with a navigator sitting behind me, and I had to learn the concept of 'crew co-operation'.

"It can be difficult until you get used to it," my Instructor had said. "Especially for ex-Lightning pilots. But good 'crew co-operation' is something you'll have to work at. Without it, the potential of the Phantom can't be fully exploited."

"Another problem," said the Instructor, "is that some of the navigators are failed pilots. It can be a bit awkward sometimes."

The pilot, sitting in the front of the F4 Phantom, flew the aircraft. The navigator occupied the rear seat, and controlled the radar system. Either crew member could be nominated as the Captain.

The Phantom had some complicated aircraft systems, including a wing-fold facility, and an arrangement called 'Boundary Layer Control'. This came into effect when the aircraft flaps were lowered; air from the engines was channelled through pipes to blow across the wings and flaps. This reduced the take-off and landing speed. "It's fine," said my Instructor, "until something goes wrong. Then there's the hazard of hot air blowing around the airframe, causing a number of potential dangers."

If that happened, a red light would flash aggressively in the cockpit: '*BLC Warning*'.

In a few year's time, I would have first-hand experience of this nightmare.

The Phantom F4.

CHAPTER 17

Phantoms & Flowers

It was about a 10 minute walk from our Married Quarter at Coningsby to the local woods. Once there, it was like being in a different world. "The military outlook seems out of place here," said Sue. "The forces of nature have taken over." A carpet of wild flowers, predominantly bluebells, was laid at our feet. Keen not to damage or disturb nature's artwork, the walker was obliged to almost tip-toe through the woods. Slow progress was inevitable.

"At least it gives us the chance to ease-up for once," commented Sue. We had the opportunity to admire the wild flowers. In the Bluebell Woods we discussed finer feelings. We contemplated a growing sense of deeper needs.

At Coningsby, we had been offered a Married Quarter. The basics of life were provided, but the lack of individuality was sometimes worrying. Also worrying was the 1970's house price boom. Sue and I discussed buying our own house in Scotland rather than relying on the Married Quarter system.

"Although its easy for you to get to work from Married Quarters, and I can have the car," she said. "I must admit it'll be nice to have our own place."

At work, the F4 Phantom course at Coningsby involved a long ground school programme initially.

"The F4's weapon system is more complex than the Lightning," said the Ground School Instructor. "The aircraft radar, for instance, has two separate facilities. Firstly, 'pulse' radar when the 'blip' (like the Lightning) shows range and bearing. Secondly, there's 'pulse Doppler'."

"The problem with the 'pulse' system," continued the Ground Instructor, "is that it's useful only at relatively short ranges. At low level, the radar range is reduced even more, because of ground returns that clutter the screen."

The F4 Phantom's 'pulse Doppler' was the solution.

"It's based on the principle of Doppler effect," said the Instructor. "For example, stand at a railway platform, and listen to the noise of an approaching train. As the train passes, the note lowers. That's the 'Doppler effect'."

"In the F4, when the Navigator selects 'pulse Doppler', the azimuth angle of the blip shows target bearing, but the distance up the screen is related to the target's speed."

"Canny," said one of the Student Navigators.

"At the bottom of the Navigator's screen," continued the Instructor, "is a line. It's a signal caused by ground returns. A blip below the line shows the target has 'negative speed'. Above the line means 'positive speed', in other words it's faster than you, or else coming towards you."

"The advantage of the system is its long range," said the Instructor. "A Navigator can detect a target at over twice the range of 'pulse'. 'Pulse Doppler' is also more effective at low levels." The students nodded wisely as they assimilated the information.

After the Ground School programme, the early flying exercises began. Then, one Friday, the Flight Commander called the course together for a briefing.

"It's time to give you guys a break," said the Flight Commander. "Next week the air-to-air combat starts."

"That's good news," said the pilots.

"Very," said a few of the navigators with a hint of sarcasm. For some of them, the air-to-air combat was a pilot-orientated activity which the navigators would not enjoy.

Over the week-end, the fuel tanks were down-loaded. Some of the dummy missiles were taken off. This was to be air-to-air combat F4 Phantom style.

"Remember the role of the F4 in the Arab-Israeli Six-day war," said the Flight Commander. "The aircraft has a phenomenal capability."

The F4, however, had to be carefully prepared for its different roles. Complex, as ever, the aircrew used a special chart which detailed the 'g' limits of the aircraft. In certain configurations, this was over 7'g'.

"You'll be crewed with Nick," the Flight Commander told me. "As you know, he's the new-boy on the block. But your Lightning flying should see you through." I looked at him doubtfully.

Nick was a gangly individual. He appeared to live in something of a dream-world. At the pre-flight briefings he seemed tense and uncomfortable. "Not sure I'm looking forward to this," he said to me.

"Don't worry, Nick, you'll be fine," I said hopefully. My fingers may have been crossed.

Nick certainly had teething problems, but after the first few sorties he suddenly seemed to 'click'.

"Target left, 9 o'clock, in a hard left turn!" Nick yelled on one flight. His head was cranked round trying to see our 'opponent'. It was bad enough for the pilot, with the 'feel' of the aircraft in his hands. It was worse for the Navigators as they were thrown around under high 'g' forces. On this occasion, Nick had become absorbed by the challenging situation. He had even loosened his top straps slightly, so that he could turn his shoulders to help with our 'look-out' to the side and rear. Maintaining visual contact with the 'enemy' was a vital part of achieving success in air-to-air combat.

"Now slipping to the 8 o'clock. But we have speed advantage," continued Nick. "Suggest high speed yo-yo..."

I made sure that both throttles were in the fully forward 'cold power' position. As the other F4 moved more towards our 'six o'clock' position, he began to gain the advantage. However, we had one chance to reverse our declining fortune. We could use our higher speed to advantage by pulling up, then banking steeply, and hopefully end up better positioned to turn in behind the other aircraft.

"Agreed," I replied to Nick's suggestion. I made sure that the airspace above was clear, and had another look at our 'opponent'. Then I called: "Standby...pulling up...NOW."

THE SPICE OF FLIGHT • 113

I hauled our aircraft into a vertical climb. The anti-'g' bladders pressed firmly against my legs as the anti-'g' suit inflated to maximum. I tensed my stomach muscles, fighting against the 'greying out' feeling which swam before my eyes. I regularly monitored the 'g' meter, keeping the needle close to the maximum limit.

As our airspeed started to reduce, I glanced back again at the other F4. The apparent size of our 'opponent' rapidly diminished as we climbed. He was attempting to follow our manoeuvre, but lacked our speed advantage. When we had reduced to the minimum safe airspeed, I banked steeply to the left.

"It's working! It's working!" Nick suddenly cried out excitedly.

"Keep turning left. Keep turning left!" Nick called. I continued to apply really high 'g', as I racked the F4 into the hard turn, and began to descend from the top point of our 'high speed yo-yo'.

"Target now coming to the 10 o'clock!" shouted Nick. "I think the sod has lost visual on us!"

This was quite something. Our 'opponents' were an experienced staff crew. The 'high speed yo-yo' technique really seemed to have worked.

"Standby...*keep turning...keep turning...*," I maintained the high 'g' loading, and engaged minimum reheats. The engine reheats were used sparingly, because of the high fuel consumption. It was a 'trade off', but at that stage I felt the 'trade' was worth the penalty, so that we could consolidate our lead.

"Try to get a radar lock," I said to Nick.

"Standby...Radar lock-on confirmed!" yelled Nick. I made some final checks around the cockpit.

"Squeezing the trigger now," I called back. The cine-camera in the front of our aircraft started to roll.

"Green 02, *FOX 2*," I then called on the radio. 'Fox 2' was a code. It meant that I had fired a simulated 'Sidewinder' missile at our 'enemy'. The call would have made our 'opponents' hearts drop.

"Green 02 from Green leader, confirm your position." I glanced at my rear-view mirror. Sure enough, I saw Nick holding a 'thumbs-up'. I almost saw the huge grin behind the oxygen mask.

"Green leader, from Green 02," I said. "We're in your 6 o'clock!"

It was the final confirmation to our 'opponent' that we had 'shot him down'.

There was a pause, then: "Well done Green 02. I guess we owe you a beer!"

Back on the ground, there were the usual lengthy de-briefs, and the staff crew were magnanimous. Young Nick was thrilled. "I can't wait to get home to tell my girl-friend!" he said.

When I got home, I felt on a high too, and told Sue about the flight. "It was quite pleasing," I said, "considering that we were such an inexperienced crew. And it was great that Nick enjoyed it so much. Some of the Navs have real problems."

"That's excellent," said Sue, "well done."

She then had a chance to tell me about her day. "Oh, I've had a good day too!" she said, and looked at me in a funny way.

My brain cells slowly began to re-adjust to more earthly matters.

"Oh, of course, you went to the Doctor today! Sorry, darling, I forgot." There was a pause.

"Well?" I enquired.

"Well what?" she asked.

"Well, about the Bluebell Woods and all that."

She got up, and walked over to where I was standing. I was still in my unglamorous Flying Suit *'Aircrew For the Use of'*. She looked at me, and gave me a hug.

"Oh, don't worry about it," she smiled and her eyes sparkled. "Those Bluebell fairies know what they are doing!"

CHAPTER 18

A New Year Arrival

The hour was approaching midnight on December 31st 1973.

The F4 course at Coningsby had been completed some six months ago. My wife's pregnancy had reached full-term, and the baby's delivery was expected the following day, New Year's Day.

I was on my own in our new house. We had bought a property just three miles from Leuchars Airbase, in Fife. The staff at the Craigtoun Maternity Hospital, near St Andrews, had just sent me home for a few hours rest before the business of the next day began to get serious.

As midnight approached, I became aware of distant New Year celebrations. There were some shouts, and the occasional fireworks were let-off. Far away bag-pipes could be heard wailing in the background.

The next morning I awoke early. My dream-world was quickly pushed aside by a surge of anticipation.

Driving along the deserted roads to Craigtoun, I hoped that I was not over-eager. I walked into the Ward, and soon saw a familiar figure standing in a Dressing Gown. The mother-to-be had been up for a while.

Just then, the Nursing Sister spotted me. "You're bright and early!" she said in her Scottish 'burr'. "I think you're in for a long day."

Sue sat on the bed in her small Ward. I read a book, and we played the occasional game of Scrabble. Even though it was New Year's Day, some workmen were felling trees. The monotonous noise of their chain-saws droned in the back-ground continuously.

Morning turned to afternoon as the long day progressed. I must have felt quite tired sitting in the peaceful Ward, just waiting. Surreptitiously, I began to nod-off. A vivid dream then took me back to an exercise which had been called at Leuchars a couple of months earlier:

The Alert siren at Leuchars had started suddenly in the night. Even though we were living away from the Airbase, the sound signalling a general call-out was heard clearly. I dragged myself out of bed. As I was getting dressed the telephone rang.

"I know," I said to the voice on the telephone, "I heard the siren from here! I'll be down as quick as poss."

At Leuchars, the well-oiled machinery was under way as the F4 Phantoms were dispersed to special shelters around the airfield. We were to be involved in a NATO exercise. I had joined Number 43 Squadron, whose emblem was the Fighting Cock. My Navigator, a tall individual with an unusual gait when he walked (hence his nickname of 'Lurch', although his real name was Bill), prepared maps, while I versed myself in current warnings and notices which might have affected our flight.

Soon we were driven to a remote part of the airfield, to wait near our aircraft. Extra fuel tanks had been bolted on to the under-side of the F4s. From Leuchars, our tasking often involved co-operation with the Royal Navy, and very long flights were typical. It was the height of the 'Cold War', and Soviet Aircraft had the habit of flying towards the North Sea to keep an eye on the Royal Navy, and to check our reaction times.

"Chequers 11 to cockpit readiness!" The loudspeaker tersely gave us instructions.

'Lurch' and I grabbed our flying head-sets, and ran out to the waiting F4. The ground power unit was fired-up, and I set the cockpit switches to receive the electrical power.

'Chequers 11 at cockpit readiness!', I called on the radio.

'Lurch' and I sat in our respective cockpits, and talked about the exercise parameters. "The Navy has so many complicated 'Missile Exclusion Zones' and other boring restrictions," said 'Lurch'.

"I know," I said. "But they'll get up-tight if we infringe them."

"Chequers 11, this is the Air Defence Controller. For Combat Air Patrol, make your initial heading 040 degrees. Scramble… scramble… scramble. Acknowledge."

I answered the radio instructions, at the same time as pressing the start buttons on the F4's engines. It was still the middle of the night as we taxied out. The dark of night was dispelled by the glow of light from our engine reheats during the take-off run.

As we climbed out to the north-east of Leuchars, we became aware of the clear and starry night above. Layers of low-level cloud lay beneath us. We levelled at 15,000 feet, and checked-in on the radio with the Air Defence Controller, based at Buchan, to the north of Aberdeen. On his large radar screen, the controller had an overview of the 'big picture'.

"Chequers 11 this is Buchan. Maintain your heading. The fleet is range 185 nautical miles." As we settled down for the long transit, 'Lurch' and I talked about the effects of our interrupted sleep pattern. Then we noticed something else. It was Autumn, and in the crisp, clear, night sky we had become aware of unusual and dramatic light patterns way up north.

"They're the 'Northern Lights'," said Bill, "amazing aren't they? It was worth getting up just to see those!" The light patterns were eerie; they would change form and colour rapidly, and bursts of light would be reflected within the cockpit. Shards of brilliant illumination shattered the surrounding darkness in random, ever-changing shapes.

"Chequers 11 from Buchan. Call the Fleet Controller now on pre-briefed frequency."

As we spoke on the radio to the Royal Naval Officer, sitting on board a ship miles below us, he took over the radar control from Buchan.

"Chequers 11 from Fleet Controller, good evening. Make your heading 360 degrees initially." We were to start a 'Combat Air Patrol' on a north-south pattern. We would be just one element of the various resources, including missiles, other ships, and other aircraft. It was the start of the exercise, and everything seemed quiet at that stage. The Northern Lights continued to entertain and fascinate us.

"Fleet Controller from Chequers 11, confirm Tanker available please", I asked. I was concerned that, even with the extra fuel tanks fitted, we would need in-flight refuelling to maintain a long Combat Air Patrol.

"Roger, Chequers 11. Tanker available in 45 minutes." 'Lurch' did a further fuel calculation, and was satisfied with the time-scale.

There were no signs of target activity, and so we asked the Controller for clearance to overfly the fleet. As we approached the few ships that comprised the fleet, a Morse signal could be seen flashing in the night sky.

"How's your Morse, Bill?" I asked.

"I think he is telling us to sod off!" said Bill. At that point, the Tanker checked in on the radio. "Good stuff," exclaimed Bill, "here comes the petrol."

We obtained permission from the Controller to head straight for the Tanker. The first signs of dawn were just peeping over the horizon. At the same time, the Northern Lights were slowly disappearing. "Show's over!" said Bill.

I called the Tanker Captain on the radio, just as Bill announced that he had radar contact. 'Lurch' gave instructions for the interception on to the Tanker, and as we rolled out behind, I could see the lights of the Tanker ahead. I operated the cockpit switch to deploy the F4's in-flight refuelling probe; it was part of the check-list read out by my Navigator.

"Tanker dead ahead range 5 miles," Bill called out. I concentrated on the heading he had given me; it was easy to get disorientated at night. Monitoring the flight instruments was important, even though I could see the Tanker visually. The small baskets trailing behind the Tanker's refuelling hoses would not be visible until we were closer.

"Tanker now range 3 miles." I did a further check around the cockpit to ensure that the switchery was correct. As it was dark, I had switched on a special lighting system to illuminate our refuelling probe.

"Tanker range approaching 1 mile," Bill said. At that point, instead of flying directly astern, I put the Tanker to one side of us so that we had a better chance of visually spotting the refuelling baskets.

"Tanker range approaching ½ mile. Confirm visual with the baskets?" enquired Bill.

"Roger. Just getting visual with the baskets now," I replied. The baskets were ringed with pale green lights. I asked the Tanker Captain for clearance to make contact.

"Roger. You are clear for refuelling," announced the Tanker Captain eventually. I eased the aircraft's throttles gently forward to close on to one of the baskets. As the airflow between the basket and F4 became disturbed, the basket was pushed away. I noticed that a faint light on the horizon was just visible; the progression of dawn helped me with a sense of orientation.

I missed the first attempt to contact the refuelling basket; some clear air turbulence had caused the basket to move about. I eased back the throttles for a second try. "Steady," Bill called from the back...

"Mr Pike! Mr Pike!" My eyes jerked open. The Nursing Sister had interrupted my dream. She looked at me crossly.

"There you are! Your wife has been taken into the Delivery Ward. You pilots, I don't know! You've all got your heads stuck in the clouds."

I quickly donned the green suit which had been issued. It made me think of '*Green Suit... Aircrew for the use of*'. I poked my head around the side of the door leading into the Delivery Room. I glanced at the wall-clock; it was 6.30pm; I had been in the building for about 12 hours.

My wife was lying on the Delivery Bed, and next to her was a Nurse carefully powdering some gloves before putting them on. My wife had a 'drip' into her arm. I went up and held her hand.

"Ready for the fray?" I asked. She nodded, and tried to smile. A spasm of pain interrupted her smile. Just at that moment, another Nurse appeared, and said to her colleague: "What about the time, Sheena? You'll miss your tea-break!" The carefully prepared gloves were pulled off rapidly, and Nurse 'B' put some on instead. Another spasm of pain from her patient made the Nurse look up quickly. "Alright, love, try to relax," she said.

I looked anxiously at my wife. "She's nearly there," announced the Nurse, "I think we'd better get the Doctor."

"Perhaps he's in tea-break," I said. The Nurse gave me a withering smile.

"Oooh...", cried my wife.

"Back in a tick!" called the Nurse.

"But don't just leave us…"

Soon the young Doctor appeared. "Ooooh…" yelled my wife again. The Doctor made an examination of his patient. "Won't be long now!" he stated.

"Oooooh…"

"Now, love. Remember those breathing lessons." The patient nodded, and then continued with her "Oooh…" conversation.

"Okay," said the Doctor eventually, "I want you to push hard… NOW." My wife obliged.

"Relax. Relax!" said the Doctor after a while. "Tell me when the next push is coming." There was a general pause in the proceedings. I looked around the room. It had a utilitarian atmosphere. Drips, and pipes, and a miscellany of medical equipment lay scattered about. I glanced at the clock on the wall as it ticked steadily. The medical team waited patiently; I looked at their green gowns and face masks.

Suddenly, I felt the grip of my wife's hand tighten. Sue looked at the Doctor and nodded. Then: "Oooooh…"

"*PUSH…*" said the Doctor.

At that moment, everything seemed to happen all at once.

The medical team, reinforced again by Nurse 'A' after her tea-break, followed a well-rehearsed routine as the baby was delivered. I continued to hold Sue's hand, and tried to stay out of the medical team's way. Nurse 'B' was wrapping the child in a small blanket.

"Congratulations!" said the Nurse at length. "You have a beautiful baby daughter."

I glanced at the clock again. The time was 7.47 pm exactly. 'Lizzie' was snuggled next to her mother.

Soviet Bear bomber under escort by a 43(F) Sqn F4 Phantom through the Iceland-Faroes Gap.

CHAPTER 19

Chasing Bears

The adrenaline had started to flow. I was on duty in the Interception Alert Force hangar, at Leuchars. We had been warned that a Soviet 'Bear' aircraft was heading towards our patch. These enormous bombers, with two propellers attached to each of the 4 engines, had the habit of performing regular 'Cuban runs'.

At the height of the Cold War, the Soviets maintained close links with Fidel Castro's Cuba. The 'Bear' aircraft sometimes flew through the Iceland/Faeroes Gap en-route to Cuba. The Interception Alert Force at Leuchars was a first line of action to intercept these Bears. Our task was mainly to show a presence, but the F4 Phantoms were fully armed, and in the extreme we were allowed to defend ourselves. Some of the Bear crews were known to have been aggressive in the past.

It was July 1974, 15 years before the collapse of the Soviet Empire. Lizzie was 6 months old already, and it was the day before my 31st Birthday. My usual Navigator ('Lurch') was away on leave, and I was crewed with Barry, an experienced man with an astute airmanship mind. Large in frame, and with a domineering personality, Barry was not the type to cross.

The long-range radars had spotted the lumbering Bear as it flew along northern Norway. Barry and I discussed the general picture, and prepared as much as possible. As it was likely to have been a long flight, a Victor Tanker aircraft had been organised.

Suddenly the loudspeaker ordered us: '*IAF 1 to cockpit readiness*'.

Barry and I strapped into our aircraft, and checked-in with the controller at Buchan. The controller gave us up-to-date information, and kept us at cockpit readiness for a few minutes. Unlike the

Lightning set-up in Germany, when the scrambles had a last-second urgency, this was a more sedate scene. The controllers had time to watch and plan the target's interception.

"*IAF… this is Buchan… for possible hostile interception… heading 330 degrees initially… scramble, scramble, scramble. Acknowledge.*"

I acknowledged the instructions, and started the F4's engines as Barry wrote down further details from Buchan. It was late-evening already as we climbed out to the north-west of Leuchars. Below us lay the familiar Fife countryside with its gently rolling hills and picturesque pattern of cornfields, still clearly visible despite the slowly fading light.

Gradually, the scenery changed to the stark and mountainous aspect of the west of Scotland. On the radio, the Buchan controller was passing information about our Victor tanker. The plan involved a fuel top-up before we set off the long distance to the Iceland/Faeroes gap. Barry performed various complex calculations which gave us options for diversion airfields. He considered different scenarios in case we had problems with in-flight refuelling.

As the tanker checked-in with Buchan, we were given vectors by the radar controller towards the refuelling aircraft. We were flying at 28,000 feet, just below the height of the airliners plying the trans-Atlantic air-corridors in a non-stop stream. Contrails criss-crossed the skies above us.

We made the planned rendezvous with the tanker and, as we sucked fuel from his tanks into ours, we continued on the north-westerly heading which took us way beyond the Scottish Western Isles, and out towards Iceland. By the time that we intercepted the Bear, it was expected to be around midnight. At that time of year, and that far north, it would not have become fully dark. Even so, the interception was anticipated in poor light conditions.

When our fuel tanks were full, we bade adieu to the Victor tanker and confirmed that we were looking forward to his services later. Meanwhile, the radar controller was giving us more information about the approaching Bear. Barry had our pulse-doppler selected, and was searching non-stop for the speed signature of our target. We flew for another 30 minutes or so towards the Iceland/Faroes gap.

Suddenly, Barry said quietly, "I think I've got him."

In order not to give our position away to eavesdroppers, we were using minimum radio. Tempting as it was, therefore, we did not ask radar for confirmation of target range. Barry, professional as ever, was still working away at the aircraft radar, and asked me to turn the F4 Phantom, in order to place the target to the edge of his radar screen.

"I'm pretty sure that's him," said Barry, and we continued to fly the new heading. Barry switched to pulse mode on the radar from time to time.

Eventually he called out, "I've got him on pulse."

A good view of a Soviet Bear Bomber on a 'Cuban run' during the Cold War.

That was good news; it meant that the blip on his screen was giving more useful information, including, crucially, an accurate indication of the target's range. In the pulse-Doppler mode, someone of Barry's expertise could have made educated guesses about the target range, but now we had it confirmed.

"Turn right 15 degrees," Barry instructed me. He was taking us on a curved pursuit, as far from the Bear as possible so that the crew would not easily see us visually. We descended a little, and then slowly began a left turn, to creep up behind our target.

"He's taking evasive action!" shouted Barry suddenly. The light was really quite poor, and there was no sign of the Bear's navigation lights; the Soviet bomber had evidently switched them off.

"Reverse turn hard right," Barry called. As I obeyed his instructions, I was straining my eyes to try to catch a glimpse of the Bear's outline. It was also important that I monitored the flight instruments; it would have been easy to have become disorientated in those conditions, which was presumably the aim of the Bear's crew.

"Hold that heading… steady… now turn hard left!" yelled Barry.

I was having to concentrate really hard. If we had lost radar contact for just a moment, the Bear would have been hard to find again, and the bomber crew knew that we were operating to fine fuel limits. The Bear had timed its arrival carefully. The bomber was doing its utmost to throw us off in this cat and mouse scenario.

However, the skill of my Navigator was paying dividends. As a crew we were pulling together really well, and we both knew that there would have been little possibility of a second chance that night.

"We're closing slowly on the target," called Barry, "Steady… standby… *Turn hard right!*"

I followed his commands, and continued to search visually for the Bear's outline.

Suddenly, magically, nature was about to help. Just on the horizon, a distant but clearly visible patch of light was appearing. I mentioned it to Barry.

"Standby; he's just rolling out from this turn," Barry said.

It was the break we needed. As I manoeuvred our aircraft to place the Soviet Bear in the clear patch in the sky, a faint outline of the bomber gradually became visible.

"Got him visually, Barry!" I called out. With Barry continuing to monitor the radar range of the target, I concentrated on watching the bomber's outline as we slowly crept up from behind and below.

The Bear continued to weave, and he made life as difficult for us as he could. But as we pulled up along side, the target became clearer as we got closer. Slowly his manoeuvres stopped. The bomber crew must have picked us up visually, and they must have realised that their evasive actions were becoming extremely dangerous to both aircraft. Barry by then was also in visual contact.

"Standby," he said, "I'll use the infra-red camera for photos."

Just as Barry was taking photographs, we simultaneously let out a cry. The rear-crew gunner had trained a powerful spotlight into our cockpit. In a split-second, he had destroyed our night vision. We had become, in effect, instantly blind. I had no alternative but to close the throttles, and to turn away, in order to create distance between us and the bomber.

I rotated my cockpit lighting rheostat to fully bright, but even so had difficulty in seeing the instruments. I went right back to basic 'Instrument Flying' techniques. A little voice kept repeating... *'Artificial Horizon... Artificial Horizon...'* To the exclusion of practically everything else, I had to concentrate on keeping the aircraft level. My first and over-riding priority was to prevent a disastrous 'unusual position' from developing. It seemed like an age, but slowly, slowly, our vision was restored.

Barry was concentrating on picking-up the target on radar again. Eventually, he called: "Target contact regained. He is on our left, slightly high, range 4.5 miles."

"Did you have any success with the infra-red photography?" I asked Barry.

He confirmed that he had taken some good shots. We did another fuel calculation, and decided that our next tactic would be to shadow the Bear in line astern for some miles, knowing that he would have been well aware of our position.

It was our duty to show the Bear that his aggressive conduct had not put us out-of-action. As we continued to shadow the bomber, I made a brief call on the radio to the Victor tanker. Using a pre-arranged code, I managed to confirm that he was still on task at the rendezvous point. We decided to continue shadowing for a few more miles.

"At least the bastard will feel uncomfortable knowing that there are eight live missiles pointing at his arse," growled Barry.

Eventually, we decided that our duty had been performed, and that it was time to break away, and to head towards our friendly tanker.

Ironically, if we had met the Bear crew face-to-face, probably we would have become firm friends quite quickly. In the anonymous environment of our separate cockpits, however, we were undoubted enemies, fighting for totally different causes.

The tanker crew, well away from the front-line, would not have had any idea of Barry's and my recent experiences. For them, it had been a long, boring, night.

Their normal, matter-of-fact voices, however, somehow touched an emotional spot for the F4 Phantom crew that evening. We discussed it briefly on our return flight to Leuchars.

It was the wee small hours as we flew over the Fife village where Lizzie and her Mum were still asleep. The lights from Leuchars airbase stood out prominently ahead.

After landing, and as we taxied our F4 back to its hangar, Barry suddenly piped up: "By the way…" he said. There was a slight pause. I was concentrating on the hand signals from a marshaller ahead.

"Well?" I asked, at length.

"Oh nothing!" he hesitated. "Just… well… Many Happy Returns!"

CHAPTER 20

A Close Call

Apparently my face was ashen-white when I returned home late that evening. I had good cause.

It was mid-way through my tour at Leuchars, which had seen mixed fortunes since the flight of the aggressive Bear. There had been a number of positive aspects, including flying as a member of a display team of four F4s at a Leuchars Battle of Britain Open Day. I had become the Squadron Instrument Rating Examiner, and in addition had taken on the job of Editor for *Contact*, the local monthly magazine for Leuchars.

On the debit side, for six months or so the Squadron had been detached to Kinloss, on the Moray Firth coast, while the Leuchars runway was resurfaced. With my wife pregnant again, it had been a difficult period. I had not got on well with the Squadron 'boss' (he was later regrettably cashiered from the Service), however, a successful Missile Practice Camp had – unexpectedly – caused something of a reconciliation with my boss.

The briefing for the flight had been straightforward enough.

"It'll be a night flying exercise, mainly for the benefit of the new pilot," said the Operations Officer.

I was tasked as the leader of a pair of F4s. Our brief was to climb to 30,000 feet or so, and to practice different interception profiles as part of the new pilot's training induction.

There had been no hint of trouble during the flight preliminaries. We had started up, taxied out, and taken-off as normal. We had levelled at the planned height, and were just setting-up for the first practice interception profile. During the climb-out from Leuchars, we had passed through some layers of thick cloud. However, at

30,000 feet, we were clear of cloud. There was some moonlight which gave quite good back-ground illumination. It probably saved our lives.

When it did happen, it was sudden, and without hint of an incipient problem.

There was a discernible 'thunk', and then everything went dead. The radios went silent. All the cockpit lighting went out. I could not speak to my Navigator. The Instruments in the cockpit all sank sadly to their resting position, and some showed 'OFF' flags. Even the back-up artificial horizon was useless.

Glancing away from this dismal scene, I concentrated on keeping the aircraft level by looking at the moon, and trying to make out the distant natural horizon.

I checked in the cockpit again, and confirmed that both engines were still working. The engine temperature gauges had stopped functioning, but the two engine revolution gauges took their readings directly from the engines. They needed no electrical power, and showed that the two engines were still turning normally.

By now the crew should have been discussing the situation. It was impossible. Our head-sets were dead.

I could not call my Navigator to read out emergency drills. Using my torch, I could have read them myself, but there seemed little point; the emergency 'flip' cards would not have covered this situation.

At that point I heard a faint cry from behind. I glanced at my mirror, and saw that the Navigator had removed his oxygen mask, and was shouting something at me.

Eventually, I understood that he was calling out: "*RAT...RAT!*"

The initials stood for 'Ram Air Turbine', although it crossed my mind that the navigator was venting his feelings. By my left elbow was a lever. When pulled down, the lever deployed a small wind-mill arrangement. It was used only in dire emergency, to give limited electrical power. But it was no good. I had already deployed the 'RAT', with no effect.

The possibility of a 'Martin Baker' let-down (i.e. use of our ejection seats) was firmly in both heads. We could not communicate, but we instinctively tightened seat-straps as much as possible. Our final recourse would have been to pull our ejection seat handles, and make the descent through cloud by parachute. It was a very real possibility. The thought of spending some hours bobbing around in the freezing North Sea, however, filled our minds with dread.

I continued to use the moon as my 'artificial horizon', and began a slow turn back towards Leuchars. Visual references outside the cockpit remained my only available means of keeping the aircraft under control. Without these references, we would have ended up in a disastrous 'unusual position'; quite quickly, I would have lost control of the aircraft.

The 'E2B' standby compass needed no electrical power to work, but electrical power was needed for its lighting. However, by then I was pointing my night flying torch at the thing so that I had rough heading information.

We were flying a version of the F4 Phantom known as FG1. It was an ex-Naval machine, and the Navy evidently liked to cut back on unnecessary batteries. It saved having to service them at sea. With the FG1, they had taken things a little too far.

Later, we discovered that we had suffered a failure at a junction point of the two main generators. Because of poor design, the RAT power was routed through the same point. All 3 of our sources of electrical power were thus denied.

The aircraft did not have any batteries at all as emergency back-up; a potentially calamitous design weakness.

By my right hand were two switches which controlled the main generators. I had tried re-cycling them several times, but to no effect. I continued to switch them on and off randomly, and slowed down to allow the other F4 to catch up.

I was unhappy about doing a close formation let-down through cloud on the new pilot's first night flight, but it had to be done if necessary. The extensive layers could be penetrated by relying on the other aircraft's instruments to take us below the cloud.

As these thoughts were occupying my mind, there was a sudden flash of light in the cockpit. One of our generators was showing signs of life. I gingerly re-set the switch once more.

Suddenly, the cockpit lights came back on. All the instruments started to erect themselves again, and I was able to speak to my navigator. It felt tenuous, nevertheless, the generator seemed to be holding on-line.

With the ability to communicate with my Navigator (and with the outside world), some rapid decision making was set in train.

We asked the radar controller for instructions to arrange a join-up between the two F4s. Then I informed everyone of our plan to follow the other aircraft down through the cloud layers. Initially, we aimed to fly in close formation. I had reservations, but saw the plan as our safest option in case the generator dropped off-line again.

After some minutes, the other F4 drew up alongside us. We were still heading towards Leuchars, and I eased our aircraft into a close formation echelon starboard position.

"It's OK to commence the descent now," I called to the other pilot.

The familiar lighting pattern of the other F4 had to be kept in just the right position. It was no easy task at night.

"Sorry this is pretty rough," I said to my Navigator. With the inexperience of the lead pilot, plus my own anxiety, the night-time close formation flying was causing some stress. It was tempting to over-control. I had to force myself to go back to basics. Repeatedly I told myself to smooth down the flying.

In the circumstances, the lead pilot was doing well. Even so, it was hard to follow him. An experienced pilot would have made the night formation flying easier.

Half-way down our descent, I noticed that the cloud had started to thin.

The action I took next was criticised subsequently by the Flight Safety Department.

However, they were not in my hot seat at the time.

As the cloud was thinning, and our generator had held on-line successfully for a while, I told the other F4 that we intended to ease out into a loose 'battle formation'. It was a less stressful position, and we still had the other aircraft in sight.

However, if we had hit thick cloud, we might have lost sight of the other aircraft. It was a 'no win' situation (as the Flight Safety

Department at least had the grace to acknowledge). The close formation had become uncomfortable, but the 'battle formation' had its own hazards.

While these thoughts were tossing around in our heads, the problem was resolved as suddenly as it started.

We broke through the cloud-base, and the welcome outline of the coast was clearly visible. The generator was still holding, and both crew members let out cries of relief as Leuchars quickly came into view. We continued to keep the other F4 in sight, and Air Traffic Control at Leuchars cleared us for a straight-in approach and landing.

After landing, the aircraft was impounded and the engineers carried out an in-depth examination.

I have looked back with anguish over this incident. An aircraft design weakness had taken us to the very edge of the abyss. The problem was precisely pin-pointed, yet it was allowed to recur...

Some 18 months later, the problem appeared to rear its head once more (although the cause was never positively determined). This time the pilot became disorientated and the F4 plunged into the sea near the Bell Rock lighthouse.

The pilot and navigator were both killed.

CHAPTER 21

Phantoms in the Fog

In just a few weeks, I would be posted away from Leuchars; my F4 flying would end. However, more turmoil was on the agenda yet.

The night-time electrical drama was still under discussion, when I was involved in another emergency situation. This time, the trouble occurred during a day-light mission.

My navigator for the problem flight was Captain Carpenter, a United States Air Force Officer on an exchange programme. We had just taken-off from Leuchars, and were climbing out to the east. A two hour flight with practise interceptions was planned.

Suddenly, there was a loud noise in our head-sets, and a red caption *'BLC Warning'* was flashing in front of my face. The aircraft Boundary Layer Control system was faulty. The flying controls, and parts of the structure were potentially exposed to hot air from the engines. Yet again, the design of the F4 – inherently complex and susceptible to fault – was laying its aircrew open to danger.

"Read me the flip-card drills!" I yelled to my Navigator.

I made an emergency radio transmission to Leuchars, and turned back towards the airfield as Captain Carpenter read-out the procedures in his American drawl. After completing the emergency drills, the crew continued to discuss the situation. Their main aim now was to put the machine down on *terra firma* as quickly as possible.

The weather was fine, and Air Traffic Control cleared us for a priority visual approach. However, there was a wind from the east that day, and an approach from the sea directly onto the runway was not possible. I had to fly inland, which took us over Balmullo Hill, before taking up an easterly heading towards the duty runway. It caused more delay, but eventually I landed the F4 close to the end of the runway. The jerk of the tail parachute as it deployed gave the crew a feeling of huge relief.

My navigator and I breathed considerably more easily as we taxied clear of the runway. I stopped, and immediately closed down the aircraft engines. Fire and emergency vehicles surrounded the aircraft as we scrambled out of our cockpits.

"Not often you get one of those suckers," remarked my American colleague.

"You're having a bad run," said the Operations Officer.

"You were lucky it happened close to home," commented the 'Boss'. "Better luck for next week's exercise." His comments certainly brought us down to earth.

Preparations for the exercise were already under way. The aircraft were made ready; Operations staff studied orders; special crew rotas were arranged; there were extensive briefings.

The exercise, as usual, involved dispersing the F4s around the airfield, to conceal the aircraft. It helped to counter the war-time threat from bombers, and also from Communist (or other) saboteurs.

After the exercise had been underway for some days, 'Lurch' and I were on duty at one of the dispersed sites. We had been waiting for hours in the basic, and uncomfortable, environment. It was around mid-day. Shortly we would be called to cockpit readiness, and the flight was about to be a marathon. I was thinking of my wife, pregnant with our second child.

"*Chequers 11 to cockpit readiness!*" It was the last time in my flying career that I heard those words directed at me.

As Bill and I ran to the waiting F4, someone handed us a meteorological warning. 'Haar', or local sea fog, was expected later in the day. "That's all we need," said Bill.

As we strapped-in, information was coming from Buchan about a Victor tanker. "Look's like we're in for a long haul," commented Bill.

As Bill was writing down details, the controller's voice came through our head-sets: "Chequers 11, make your heading 035 degrees ...*scramble, scramble, scramble!*" I pressed the start buttons of the F4, and the aircraft's nose-wheel steering facility proved useful as I manoeuvred away from the dispersed 'hide-out'.

We both felt glad as the aircraft climbed away to the north-east of Leuchars. It was a relief to escape the tiresome waiting around. The weather was clear at that stage, and further north along the coast, the outline of Aberdeen was becoming visible. The North Sea Oil was in its infancy then, and occasionally we flew over helicopters making their way to the Forties Field, and other new installations. Further out to sea, we noted some ominous signs of shallow fog. With the south-easterly winds, the conditions were ideal for blowing the 'haar' on-shore.

"Chequers 11 we have a target for you!"

The Buchan controller had timed our scramble well. "Turn right now, heading 085 degrees."

As I obeyed the controller's instructions, Bill was searching on the pulse Doppler radar mode. Suddenly, he let out a cry: "bloody hell!"

I waited for him to expand the conversation. Nothing was forthcoming, and so I prompted him. "Yes, Bill?"

"Standby," he replied, "the signal is still weak, but I think we have a high speed target. He's right at the top of the screen. It looks like he's mega-fast." I told Buchan that we had the target on radar, and that he appeared to be high speed.

"Roger, Chequers 11, that's affirmative. You are tasked to visually identify." There was a slight grunt from the cockpit behind me.

"Better make sure the tanker's handy," said Bill. "We're going to use an awful lot of fuel on this one."

Bill had started searching with our radar's pulse mode. With high speed targets, the Navigator's skill was really put to the test. If we rolled out too far behind, it could be a long, fuel-consuming, haul to catch the target. If we were too premature, there was the danger that we could roll out in front of the target.

"Accelerate to Mach 1.3," ordered Bill.

I was not expecting Bill to demand supersonic flight so early, but it was no time to argue.

"I've got him on pulse," said Bill. "Give me Mach 1.5."

That command surprised me. It was quite unusual. I had already selected the reheats on both engines. I pushed both throttles fully forward. The ramps by the engine air intakes would have retracted to their full extent.

"You've got full power, Bill," I assured him. I watched the F4's Machmeter. It started to increase: Mach 1.3... Mach 1.35... Mach 1.4. I also kept a weather eye on the fuel gauges as the reheats relentlessly gobbled-up reserves. It was an alarming sight.

"Turn right 25 degrees," said Bill. I had to apply a high angle of bank in the turn to meet with Bill's orders. At least in supersonic flight there was no buffet in the turns, although unwanted height inaccuracies were easily induced by rough flying.

"Standby to reverse the turn," commanded Bill. "The target's high speed still, coming down our port side... standby... *turn left now*"

As I pulled the F4 into a hard left turn, I kept both throttles fully forward to maintain Mach 1.5.

"Target now crossing the nose range four miles," exclaimed Bill.

By then, I had been catching visual glimpses of our target.

"Keep turning hard," said Bill. "We should roll out two miles behind."

I had to hand it to Bill: to have arranged our interception so accurately was highly skilful.

"Range three miles; keep turning," said Bill. Clear skies were helping us, and the target was still visible as we continued our hard turn. "Target coming into the 12 o'clock range two miles," continued Bill.

"Well done Bill," I said. "Closing visually to confirm identification."

I had already noted that our target was a high speed Lightning aircraft. As I flew alongside, the Lightning pilot rocked his wings a couple of times. "One of your old mates!" said Bill.

As we turned away from the target, Bill gave me a heading towards the Victor tanker.

I re-checked our fuel, and we made further calculations. We satisfied ourselves once more that there was sufficient fuel in the tanks to reach a diversion airfield if needed. Then we both instinctively wound-down after the concentration of the high-speed intercept.

During the transit to the tanker, Bill made comment about my impending change in lifestyle. My posting notice had been received: I was to train at the Central Flying School to become a Flying Instructor.

"Just think of the bullshit!" groaned Bill.

"You've prepared me well, Bill."

"Yeah, yeah, yeah!"

Our banter was interrupted by the radar controller: "Chequers 11 this is Buchan. The tanker will turn shortly, suggest you take up a northerly heading."

As we followed the controller's directions, there was a spell of silence between the crew.

After a while, Bill asked me: "Sue's expecting quite soon, isn't she?"

"Could be any day now', I replied.

"Hope it's not today!" said Bill.

After we had topped-up our fuel tanks from the Victor tanker, we had been airborne for less than an hour.

"Any more targets for us?" I asked the Buchan controller.

"Negative," replied Buchan. "Set up a Combat Air Patrol in your present position. Head 270 degrees initially."

Bill continued to monitor the aircraft radar as we flew around the Combat Air Patrol pattern. He found no targets, and after a while complained of feeling soporific.

"After the long hours of this exercise, we all feel the same way," I said. Every so often, I was also having to fight to keep wide-awake.

From time to time, we spoke to Buchan, but there was no 'trade' for us. The hours ticked by slowly.

Meanwhile, we were getting reports of the weather conditions at Leuchars. As the 'haar' was rolling off the sea towards the airfield, the visibility reports were becoming worse.

"It looks unlikely we'll be spending the night at home," remarked Bill.

Finally, we heard a message from the radar controller: "Chequers 11 this is Buchan. The visibility at Leuchars is deteriorating rapidly.

Return to base immediately. Be advised, the Edinburgh weather is good if needed for diversion."

I turned towards Leuchars, and began to set-up the cockpit switches in preparation for an approach using the Instrument Landing System, the ILS. Meanwhile, Bill had checked our fuel calculations again, and confirmed that we had sufficient for the diversion to Edinburgh, if it became necessary. "Better make this a good 'ILS', Mister Instrument Rating Examiner!" Bill suggested.

Buchan handed us over to the radar controller at Leuchars. We were directed to the start point of our ILS.

"Are you still awake, Bill?" I asked.

"With your flying, always!" came the reply.

"OK. OK. I hope you've got your diversion kit," I said.

"I've got a toothbrush!" confirmed Bill.

"That's a relief," I said.

"Chequers 11 this is Leuchars radar. You have 10 miles to touchdown. Turn right now, head 245 degrees. Call me contact with the localiser."

Concentrating on the artificial horizon as the primary instrument, I applied 30 degrees angle of bank to turn towards the required heading. After a short time flying in the direction of 245 degrees, the localiser needle on the ILS display started to move towards the centre; I applied bank again to turn towards the runway heading of 270 degrees. When the needle became central, it showed that we were in line with the runway. I watched any movement trend on the needle, and then used the aircraft compass to make small heading adjustments. I avoided 'chasing' the needle, which led to over-controlling.

"Localiser contact," I called the radar controller.

"Roger. Call descending on the glidepath," replied the controller.

By that stage, Bill had read out the final approach checks. Our wheels were down. Our flaps were selected. The Boundary Layer Control system seemed to be working this time; the system was puffing air happily over our flaps. There was no more banter from Bill.

The glidepath needle started to move down from the top of the ILS display.

"Chequers 11, contact with the glidepath," I called the radar controller.

"Roger, Chequers 11. You are cleared for the ILS. In the event of a missed approach, use standard procedure."

I had to keep both needles in the centre of the ILS display. In spite of my sense of fatigue, the concentration needed for this ILS was intense. As I made minor corrections, I was aware that we were flying in thick fog. I had turned off the landing lamps to prevent glare; the fog acted as a reflector.

Suddenly Bill called: "Height approaching 500 feet."

We had just a short time before reaching our 'Decision Height'. At that point, if we were still in fog, I had to over-shoot.

"Height check 400 feet, " Bill said. There was an unusual edge to his voice.

Fortunately, the flying conditions were smooth; there was hardly any wind in the foggy atmosphere.

"Height check 300 feet," Bill called out.

We had just 50 feet to go before our 'Decision Height'. The 'ILS' had been flown accurately; the needles were neatly crossed in the centre of the display. However, we were still in thick fog.

"Height check 270 feet!" Bill prompted me.

We had only 20 feet to go to 'Decision Height'. My hands anticipated pushing the throttles forward for the over-shoot.

At that second, I caught a faint glimpse of a runway approach light.

Suddenly, just at the point of 'Decision Height', I saw the runway approach lights emerging from the fog.

"Visual," I said to Bill

"Well done," he replied.

I called the Radar Controller: "Chequers 11. We have the runway lights."

"Roger. You have been cleared to land," replied the Controller.

As we touched down on the Leuchars runway, I asked Bill if he was OK.

Bill grunted a reply. He sounded very fatigued.

I taxied the F4 back to its dispersed site, aware of the swirling fog forming patchily around the airfield. Bill and I then finally, stiffly, climbed down from our cockpits. As I signed-in the aircraft at the dispersed site, the telephone rang.

"It's for yoo-hoo," said another pilot. I took the telephone receiver.

"How long was that flight?" It was the Squadron boss. I told him that the flight had lasted 7 hours and 15 minutes exactly.

"Blimey!" he said. "You must be in need of a pee!"

Bill was slumped into a battered armchair, half asleep.

"You'd better go home," the boss said to me. "By the way," there was a slight hesitation in his voice, "I spoke to Sue," he continued. "I warned her you may not get back because of the fog." There was an ominous moment of quiet. "You scraped-in just in time." He faltered again. It was uncharacteristic. "I got the feeling you should hurry," he said eventually.

CHAPTER 22

Up with the Arrows

The familiar drive home had an unfamiliar feeling that night. My tired mind had started to race; the difficulties ahead seemed to be exaggerated by the sense of fatigue.

Our house seemed quiet as I rather nervously turned the key. Lizzie, nearly two years old by then, came bouncing up to me. She flung herself into my arms, and babbled endless nonsense. That was normal; I was used to it.

I looked enquiringly at my wife. "You seem shattered!" she announced. I told her about the long flight. "I didn't think you were going to make it home with all the fog," she said. I felt it best not to tell her what a near thing it had been.

"We'd better feed and water you," said Sue. I told her to relax. I was more concerned about her state than my own.

"Oh, I'll be fine," she said, stalwart as ever.

However, as the evening wore on, it became clear that we might be in for a night without sleep. The signs and tweaks affecting my wife could not be ignored, much as we would have wished. We were both very tired. Finally, it became clear that action had to be taken. Lizzie was fast asleep when I rang the Craigtoun Maternity Hospital.

"Ach well, Mr Pike. You'd better bring her in straight away!" the voice sounded familiar from our previous visit.

A hurried telephone call was made to Mary, our reliable neighbour: "Don't you worry. I'll look after Lizzie." It was 10.40 pm as we drove off.

The journey to Craigtoun did not take long.

Unlike my F4 flying, just a few hours earlier, my eyes were not inclined to droop. I was wide awake. I helped Sue as we hurriedly climbed the steps at the entrance to the Craigtoun Maternity Hospital. We soon came across the Nursing Sister, brisk and experienced, who we recognised from last time. She took one look at Sue, and promptly wheeled her into a side-ward.

My wife stayed in the side ward for an hour or so. Shortly after midnight, however, she was taken to the delivery room. After a while, the duty Doctor was summoned. Like me, he appeared to have been a victim of sleep deprivation. He was careless, and cut the top of the baby's head while examining the mother. He caused a scar, still visible to this day.

As the magical and mysterious process of birth took its course, I held my wife's hand.

The delivery went well apart from one unusual aspect: when the baby was born, its face was upwards. For a brief moment in time, I had the strange sensation of watching a living (but unbreathing) face of a male or female person. The person was my son. When the delivery was completed, I glanced at the clock. 'Alan' had been born at 5.55 am. It was October 1st, 1975.

When she was recovering after the birth, I said to Sue at one point: "I haven't had any rest for over 24 hours." She didn't appear overly sympathetic at that stage. "Just a minor detail," I said, somewhat deflated.

During the next month, as well as having to cope with the new baby, we prepared for our move to Little Rissington in Gloucestershire. We would spend 6 months there, while I attended the Central Flying School of the Royal Air Force. The Central Flying School had a reputation for its high standards in training new Flying Instructors.

The planning for our move to Little Rissington was a logistical masterpiece. There were tears as we left our cherished first house. "Goodbye. Good luck!" Mary, and other staunch neighbours, waved as we drove off our two cars in convoy. The car I drove, an old banger, was susceptible to breaking down in wet conditions. It was tested to the limit that day. Fortunately, we interrupted our journey half-way by staying in Sedbergh, Yorkshire, with Angela and Colin, Sue's sister and her husband.

The following day, the dubious vehicle made it to our Married Quarter at Little Rissington, but only just. Despite efforts over the next few days to fix the car, we eventually decided to conduct its final journey: to the local scrap yard.

My course at the Central Flying School soon got under way. The initial weeks were spent in the Ground School. Practise was given in speaking, and teaching, in front of a group. We learnt deeper levels of aerodynamic theory, meteorology, and technical aspects. We also learnt about fault analysis, and some of the psychological considerations when dealing with students. We discussed techniques in trying to work out whether a student might have had a temporary problem, or whether he was just unsuitable as a pilot.

As an ex-fast jet pilot, I had been earmarked as an Instructor on the Gnat advanced trainer. Once the course had been completed, it meant a posting to Valley in Wales.

Our flying as trainee Instructors on the Gnat was at Kemble, about 10 miles from Little Rissington. The Jet Provost and Bulldog trainee instructors flew at Little Rissington airfield.

"The system used at the Central Flying School," said the Staff Instructor, "is for the Trainee Instructor to fly a teaching exercise, for example 'effects of controls', with a Staff Instructor. The exercise is known as 'give'. The aim is for the Staff Instructor to teach the exercise in an ideal manner. You'll then fly with a fellow student, and take it in turns to teach that exercise to each other. The next stage is for the student to 'give-back'; you'll teach the exercise to a Staff Instructor. When that's been completed satisfactorily, you'll progress to the next teaching exercise."

When we started the flying at Kemble, the students noted a relaxed atmosphere. Many of the Staff Instructors were ex-fighter pilots, with similar back-grounds to some of the students. We consequently felt quite at home.

One of the Staff Instructors I had known in the Lightning world. Lloyd (nicknamed 'Lou') had a lilting West-country accent, and his eyes would disappear in creases when he was amused. Another Instructor, Dusty Rhodes, I had known in the Hawker Hunter world at Chivenor. On occasions the student and staff discussions got high-flown, even quite heated. It was invaluable having people like Lloyd and Dusty, with their down-to-earth horse-sense combined with a wealth of experience.

As well as the instructional side, Kemble was also at that time the home-base of the Royal Air Force aerobatic team, the 'Red Arrows'. In spite of the easy-going atmosphere, there was nevertheless a strict 'them and us' policy. It was not that they were unfriendly, but the Red Arrows team required a certain ego-boost.

However, occasionally we were invited to fly with the Red Arrows. I had the opportunity twice.

In the back seat of the small Gnat aircraft, on my first flight with the Red Arrows, I was a passenger flying with the pilot in the number '5' position. It was the last day of March, 1976.

The formation leader called on the radio: "Red Arrows check-in!" The response was a rapid count: *"Two." "Three." Four."*…up to nine; usually there were nine aircraft for the practices.

The formation leader then called Air Traffic Control for clearance to taxi out. The ground crew were neatly lined up to marshal each individual aircraft. It was part of the act.

On the runway, all nine aircraft squeezed into position. We took off in two waves, with a short space between each wave. It allowed just enough room in case one aircraft had to abort take-off in the event of a problem. Once airborne, the two waves joined up quickly, and the formation leader flew to a training area.

The Gnat aircraft was ideal for the Red Arrows. Highly responsive, it provided the quick reaction needed for the Red Arrows-style precision flying. At position number 5, we were in the middle of the formation. I was used to formation flying, but this was quite different.

The few radio calls made were clipped, and sometimes coded. Many of the manoeuvres were made without radio calls. The team had exhaustively briefed and rehearsed their routine. Sitting in the back cockpit, I had been asked to say nothing to the pilot. He needed 100% concentration.

When the leader pulled up for a loop, I was struck by the vigour with which my pilot maintained his position in the formation. He did not over-control, like the beginner, but he was still rougher with the controls than I had expected. The slightest move away from his required position was instantly, almost ruthlessly, corrected.

As we pulled over the top of the loop, and the Gloucestershire countryside began to fill the view in our cockpit canopies, I noticed the other aircraft using the same technique.

When the formation came out of the loop, pointing vertically downwards, the leader made a radio call.

Immediately, some of the Gnats turned rapidly through 90 degrees, and began to fly away from the main formation. Our aircraft was one of the detached Gnats.

We seemed to have departed from the main formation so quickly, that it felt almost surreal as we suddenly flew alone at ultra-low level.

At exactly the right moment, however, my pilot pulled up and turned back towards the main formation. In the hard turn, pulling 6'g', the main formation soon came back into sight. The pilot increased our speed to 400 knots as he raced back to re-join the main formation, which by then was in a banked turn.

Quite quickly, the other detached aircraft became visible to us again.

Two of them remained detached, and went through their own routine. These aircraft pointed at each other, with one Gnat 'rolling' around the other as they passed as close as possible (this exercise was banned at a future date after a fatal accident).

Meanwhile, the remaining two detached aircraft (including ours) re-joined the main formation which continued flying a banked turn, allowing space for the two separated Gnats to operate.

When the detached aircraft had completed their programme, they joined up with the rest of the Red Arrows. The leader then progressively tightened his turn. When satisfied with the position of the formation, the leader rolled out from the turn. He then pulled up for a further loop.

When we were pointing vertically upwards, the Red Arrows' leader called a formation change.

I had been warned in the briefing. Even so, I was caught by surprise. Two Gnats, including number '5', made the formation change simultaneously. The move was so swift and slick that it must have looked impressive to spectators. However, sitting in the back cockpit of number '5', the movement was so violent that it banged my head against the side of the canopy.

The Red Arrows continued their loop, in the new formation pattern. The leader then went into further elaborate manoeuvres. The team were experimenting with new formation ideas.

Eventually, the leader was positioning his team for the finale. Speed was increased, and the formation started a vertical climb. Half-way up this climb, the leader called: '*Outwards…GO!*' The Gnats turned outwards, and the red, white, and blue display smoke emitted from the aircraft formed a spectacular shape when viewed from the ground.

As the routine drew to a close, the Red Arrows leader spoke to Air Traffic Control for permission to re-join at Kemble.

The formation approached the Airfield, at low level, and the leader called: *"breaking …now!"* Each Gnat pulled up in sequence, and turned downwind.

The aircraft then landed, and were marshalled-in by the smart ground-crew. As they walked from their aircraft, the pilots chatted amongst themselves before gathering in their briefing room.

There were two passengers flying with the team that day. We were not invited to attend the Red Arrows de-briefing; these were kept strictly in-house.

Leaving the Red Arrows part of the building, I walked from 'them' to 'us'. I bumped into Dusty Rhodes.

"How did it go?" he asked.

I told him about the head-banging manoeuvre.

"That's a bit rough," he said. Then he looked at me thoughtfully; one eyebrow was slightly raised.

"Although," he said, "I suppose we've got to knock sense into you somehow!"

CHAPTER 23

Tragedy Strikes

It made the blood run cold. It was one of those tragic situations when it was almost impossible to have avoided saying… '*if only…*' It was a last minute change, made by the out-going Squadron Commander himself, which meant that I was not part of the doomed formation flight.

I had just finished the course at the Central Flying School, and in the Spring of 1976 had driven to Valley in Anglesey with my family. We had moved into a Married Quarter at Valley, and it was one of the first days in my new job.

"No offence", the Squadron Commander had said to me, "but as you have only just arrived, I think it would be more appropriate for some of the old hands to fly in my *grand finale*."

"That's OK," I had replied, although I was rather disappointed as the flight sounded fun.

The decision may have saved my life. On the other hand, if I had been flying, perhaps I could have influenced the events.

We will never understand fully what went wrong. The first we at Valley knew of the disaster, was when the Crash Alarm sounded. It was such a melancholy noise. The Mountain Rescue Team was ordered to rendezvous. At that stage we were not sure, in our Squadron Operations, what had happened. But we had our suspicions.

The planning, and briefing for the flight had seemed normal. Three Gnat aircraft were scheduled to have flown an exercise in the low flying area of central Wales. The aircraft were crewed by Instructors; it was a 'Staff Continuation Training' exercise. The outgoing Squadron Commander was the formation leader. After the briefing, when one of the 'old hands' suddenly became available, the leader decided on the crew change.

The initial part of the flight had evidently gone well. After departing from Valley, the formation leader took the three aircraft into the low flying area. Two of the Gnats were in a 'loose echelon' formation, with the third aircraft line abreast, holding a type of 'Battle Formation' position. At a turning point, two of the aircraft initiated a formation change.

At that point, someone spotted an F4 Phantom aircraft in the low flying area. The position of the F4 was duly called. All eyes would have swung in that direction. Everything up until then had been done correctly; it was a situation for which we trained regularly.

Except on that day, something went terribly, horribly wrong. Perhaps there was a further distraction. Maybe a cockpit emergency arose at a crucial moment.

Three Gnat aircraft had taken-off from Valley on that Spring morning in 1976. The lead aircraft returned alone. The other two Gnats had collided with each other. All 4 Instructors were killed.

My wife, and a number of other wives and relatives, heard a newsflash on the BBC. It was some hours before their minds were put at rest. For 4 families, however, it was a different matter.

From our Married Quarter at Valley, we saw the lights in some of the other houses burning brightly during the following night. Just over the road from us, a young American wife, widowed that afternoon, talked with friends throughout the long and sombre night. We had not even met her, as our family had just moved in, although I had been introduced to her husband, Captain Keiffer, at work. She was flown home by the United States Air Force the next day. We were never to meet her.

In the local Chapel at Valley, there was a combined funeral service. The Chapel was packed full as wives, parents, and other relatives mingled with the staff and students present at the service. The tragic atmosphere was almost unbearable. There was a sense of disbelief. Muffled sobs dashed the aura of silence. Some people, including the Squadron Commander, were so devastated that they had to leave the Chapel and wait outside.

When the funeral service was over, and other formalities had been dealt with, it was our duty to try to return to some form of normality, in spite of the difficulties. From a purely practical point of view, half of our complement of Flying Instructors had been wiped out. With so few remaining Flying Instructors, we had a problem coping with the students. Nevertheless, as a tribute to those lost, apart from anything else, we were aware of the need to progress the student programme. There was a deadline which still had to be met.

The remaining Instructors all felt that it was necessary to ensure that everyone, staff and student, had a flight as quickly as practicable. It was a way, we had found from past experience, to have broken down any 'barriers' created by such a traumatic event.

From my own point of view, it was important to start applying the techniques just learnt at the Central Flying School. This time, however, it was with the challenge of real-life students.

If a student was struggling, success was potentially more rewarding for the Instructor.

I had a particular student who was having problems. He was an ex-manager from Sainsbury's, and was determined to pursue a flying career. He had paid for private flying lessons, and had a number of flying hours under his belt. He had persuaded the Aircrew Selection Board to give him a chance. He was really keen, which made it all the harder for the Instructors when it became clear that he was unlikely to make the grade.

"I have control," I said, "I'll demonstrate another circuit. Follow me through lightly on the controls."

"You have control," he said.

As we turned down-wind, I went through the 'patter' routine of an Instructor, describing what I was doing, and trying to put it in a way that would get through to that particular individual.

"Note the heading accurately. I'm flying parallel with the runway, making allowance for any cross-wind. Now I will do the down-wind checks, and call Air Traffic Control. It's important to fly an accurate height."

"As we get level with the touch-down point, at the end of the down-wind leg, I'll start the 'finals' turn. Note the angle of bank we're using for today's wind conditions. As we roll into the turn, I'll start the descent and call 'finals' to Air Traffic Control." I tried to make my own flying smooth and accurate, and regularly emphasised the importance of a good 'look-out' for other aircraft.

"Now I'll do the 'finals' checks. Note again the touch-down position at the end of the runway, and aim for it. As we descend, I'm gradually reducing our speed. Now I'm rolling out of the 'finals' turn, and pointing at the touch-down position. Remember what I said about flying the speed accurately!"

Sometimes, I would have introduced exaggerated throttle movements at that stage, to have demonstrated the kind of problems that could have resulted from bad throttle management. With this particular student, however, I felt it was best to avoid anything other than the 'ideal'.

"As we come over the runway threshold, note our height. Just before the wheels touch the runway, I'll start a 'round out', to cushion our touch-down. Note the throttle position. Once the wheels have made contact with the runway, I'll keep the nose-wheel in the air, and open the throttle to full power for the go-around for another circuit."

I pushed the throttle fully forward, and during the go-around, I asked the student if he had any questions.

"No thanks," he said, "that seems fine."

"Great. You have another go then." I paused a moment, then said to him: "You have control."

"I have control," he replied.

The student took over the flying controls of the Gnat and started his turn onto the down-wind leg of the circuit. "That's a good heading," I said. "Roll out there."

As we continued down-wind, I said: "Watch that aircraft ahead. You will have to arrange our spacing accordingly." That seemed to fluster him, and the Gnat started to lose height. As he corrected the height loss, I sensed a slightly panicky feeling from the poor student. The heading started to wander. The speed became inaccurate. He called out the down-wind checks, but they were incomplete, and he sounded uncertain.

"Try to relax," I said, "you're doing fine." That appeared to reassure the student. However, at the end of the down-wind leg, he tried to turn in too early. We would have ended up dangerously close to the Gnat ahead of us. The student hesitated, and the Gnat started to wander. We were flying well off track. It was a dilemma: if I took control, it would have undermined his confidence. However, if the situation was allowed to develop too far, it would have become hazardous, and caused even more upset.

"Tell you what," I said. "Let's clear the circuit for 5 minutes."

I called Air Traffic Control, and told them we would depart the circuit area to the north. Air Traffic Control were unfazed; they were used to the problems of students.

We flew away from Valley, towards the Menai Straits, with the Snowdonia mountain range providing a dramatic back-drop. As we flew along the coastline of Anglesey, small sailing craft were taking advantage of the breezy conditions, and triangles of colour bobbed along in the choppy sea. Reflected sun danced with the waves as they surged onshore. The scene seemed to have relaxed my student.

"Let's start a turn back towards Valley," I said eventually.

I made my student call Air Traffic Control. He was not over-loaded at that point, and I felt that it was good psychology for him to have managed as much as possible of the flight.

"Roger, you are clear to rejoin," said Air Traffic Control. "Be advised, there are two other Gnat aircraft in the circuit." That information unsettled my student. As we joined the circuit, and started the down-wind checks, he was unsure and faltering. At the end of the down-wind leg, for the second time he failed to turn at the appropriate place.

"Are you OK?" I asked.

There was a pause, then he said: "Would you mind taking control, please?"

"I have control," I said, and turned the Gnat to the 'dead-side' of the circuit to give us some breathing space.

"Sorry about that," he said. "But I think if I had just carried on, I would have killed us both."

His words shattered me.

CHAPTER 24

A Cloud on the Horizon

It was my first student failure. The fact that my student had shown such commitment just rubbed salt into the wound.

My next-door neighbour in the Married Quarters was a fellow Flying Instructor, and I decided to talk the situation through with him. Fairly short in stature, and rather intense by nature, Al, my next-door neighbour, was inclined to mood swings. "Al and his wife are dying for a family, and things are not working out for them," someone had told me. They would look wistfully at our boisterous youngsters causing chaos in the area. Alan had just learned to walk, and Al nicknamed him 'Douglas Bader' as the toddler stumped around stiffly. Despite some hang-ups, Al was an experienced Instructor and I found it helpful when we had discussions about student problems. We sat outside in our garden one weekend.

"I can't help the feeling that I didn't do enough," I said to Al.

"Some students, no matter how keen, just haven't got the aptitude," he replied.

"Then how did this one get so far in the training process?"

"The students reach a plateau. Sometimes you can see it coming, at other times it happens unexpectedly. It also works the other way round, with the slow starter suddenly 'clicking'." I thought back to my own student days. It concerned me even more, because in my case an Instructor change had stimulated the improvement.

"Don't get uptight about it," said Al. "And bear in mind that students are pretty unpredictable in general."

"That's true!" I said.

"However much we may feel sorry for an individual," continued Al, "he must reach a minimum standard, otherwise he's a danger to himself and others. Border-line cases are naturally the difficulty. There's an argument that by failing doubtful students, we err on the safe side, and anyway our system expects a fairly high percentage rate of failures. Other training regimes, the Israeli Air Force, for example, have a different policy. They'll reject something like 90% of potential pilots at their aircrew selection process, accepting the loss of slow starters. They'll then concentrate more resources into ensuring a high pass rate for those selected. They reckon it's more cost effective."

"So you think that my student would have failed to make the grade anyway?" I asked.

"We can't be 100% sure, but that student reached a plateau early in the course, and it begged the question whether it was right for more resources to be spent on him. Think of his final remark on your last flight with him. Sure, it was spur of the moment, but it showed that his under-lying lack of confidence had reached the stage where the resources needed in such a case would have become disproportionate."

I still had the feeling that I should have done more to build up the student's confidence. It was perhaps hard as a new Instructor to have seen the big picture. However, I was conscious that the class-room discussions at the Central Flying School about 'student psychology' had offered no more than a few basic ideas. Dealing with real cases was a different predicament.

At that moment, I felt a thump on the back of my head. I turned round with a sense of annoyance at the sudden interruption to our musings. 'Douglas Bader' came charging up with a grin that stretched wide. Al appeared to stifle a smile.

"Hi," he said to 'Douglas Bader'.

'Douglas Bader', however, was interested only in recovering his plastic ball. As he went to recover the ball, a loud squeal from his sister caused a further diversion, and 'Douglas Bader' stomped off to investigate.

"Oh well," sighed Al as 'Douglas Bader' disappeared again, "at least we've got an interesting flying programme lined up for next week."

The formation flying phase had been planned to start that week.

The formations usually consisted of 3 Gnat aircraft, with an Instructor flying the lead Gnat, and two students following. These two students flew with an Instructor for the first few flights, then they were cleared for solo flying.

"Green, check-in," the formation leader would call.

"Green 2."

"Green 3."

The formation leader then spoke to Air Traffic Control, and the 3 Gnat aircraft taxied on to the runway for take-off. "Rolling…now," called the leader, as he released the brakes at the start of the formation take-off.

As the formation climbed to operating height, the students had a chance to get used to the correct formation position. If an Instructor was in the Gnat, he gave guidance, and if necessary took control of the aircraft to demonstrate the best techniques. The leader made the turns and wing-overs gradually steeper and more demanding. When he was satisfied with the standard of his 'wing-men', the leader called: "standby for barrel roll!"

The leader then turned and dived slightly to start the 'barrel roll'. He tried to keep throttle movements to a minimum so the 'wing-men' were not encouraged to over-control.

"Pulling up now," called the leader, as he progressively applied 'g' to raise the nose of his aircraft above the horizon. The leader continued to roll the formation through the inverted, before returning to level flight, aiming to end up at the speed and height at which he started.

When the students were at a more proficient stage, the leader had no need to make these calls before starting a manoeuvre. He would have gone from the barrel roll, for example, directly into a loop. With the leader pulling 4 'g' around a loop, it was most satisfying for an Instructor to watch a good student maintain an accurate formation position. Many of our students were highly talented. In 6 years time, some of these students were to become operational Harrier pilots in the Falklands War.

Interspersed with the formation flying, we practised an exercise which involved climbing to high level, and then descending to low level as quickly as possible. That exercise ultimately led to the end of my fixed-wing flying career.

"We're level now at 30,000 feet," I said to a student, "so I'll demonstrate a maximum rate of descent. Ready?"

"OK. I'm ready."

"The best procedure to enter the max-rate descent is to roll the aircraft inverted, and pull the nose down about 30 degrees." It was a slick manoeuvre, and saved an uncomfortable 'bunt', or push, of the nose of the aircraft.

"The Rate of Descent Indicator's 'off the clock' now," I said. "but we're descending at a rate of around 15,000 feet per minute." The two minute descent from high level was taxing on the sinuses, especially if the pilot had a slight cold. After the Instructor's demonstration, the student took control to climb back up to 30,000 feet.

"Run through the pre-descent checks," I said, "then have a shot at the max-rate descent profile."

As the student rolled the Gnat inverted, and settled into the steep nose-down attitude, I felt my sinuses tweaking. I watched the aircraft instruments; the altimeter was winding down so rapidly that it looked a blur. Suddenly, I was aware of a lot of pain. I briefly removed my oxygen mask, and realised that I had a nose-bleed. "Standby!" I said to the student. "There's a problem here. Just level off at this height please." The student eased the nose up as he levelled the aircraft.

"Sorry about that," I said. "We'd better return to base."

"Are you OK for a slow descent?" he asked.

"Just take it steady, and we'll see how it goes." The student started a slow descent as he turned towards Valley, and spoke to Air Traffic Control. After landing, the Squadron Commander had said: "You'd better see the Doc straight away."

"I'll give you drugs to dry up the sinuses for now," said the Doctor. "But, Richard, I think you realise this is a long-term problem. The operation you had 10 years ago helped a bit, but further surgery isn't a useful option. I'm afraid if this carries on, there'll be no choice but to ground you." His prognosis, although not entirely unexpected, sent a chill through me.

CHAPTER 25

Farewell to Jets

After a few days on medication, I had been allowed to return to flying. At the back of my mind was the constant niggling thought that a drastic change to my flying career was just around the corner.

"Would you mind doing an Air Test?", I was asked by the Operations Officer shortly after returning to flying duties. It was the 17th of November, 1976, the year when Concorde began its supersonic service across the Atlantic.

"Sure. No problem." I had replied.

"The engineers have adjusted the fuel control system on the engine of Gnat '20'. You'll be flying with Dougie on a quick Air Test." Dougie was the Unit Test Pilot. He was not a Flying Instructor, and had a slightly vacant air. He had been at Valley a long time. "He doesn't seem to fly much, or keep in current practice," said a colleague one time. "Makes you wonder how he'd cope with an airborne emergency."

On the Test Flight, it was my job to note down temperatures, pressures, and other parameters. "We'll climb up to 30,000 feet to do some slam throttle checks," said Dougie. He had just returned from leave, and seemed even more distracted than normal.

As the Gnat ascended, Dougie and I remarked on the good weather conditions. We flew towards the Lleyn Peninsular, and vividly silhouetted below us we saw the outline of Anglesey, and the North Wales coast. Above us was a prominent stream of high-level contrails pointing out towards the Atlantic. "Perhaps one of them's Concorde," said Dougie. We heard other Gnat pilots talking on the radio. There was a busy training programme, as the aircraft took advantage of the fine weather conditions.

"Just approaching 30,000 feet," said Dougie. "Ready to write down the figures?"

"Standby," I said. "OK. Ready when you are."

"Roger. Closing the throttle." There was a pause. "Slamming the throttle fully forward...*now!*" said Dougie.

There was a bang, followed by a declining engine noise.

The engine started to wind down quickly. The engine compressor-speed gauge fell to zero. The engine temperature gauge reduced rapidly. The inside of our canopy iced-over promptly; the heating system had failed with the engine's demise. It amazed me how soon we began to ice-up. In spite of the fine weather, we were flying blind, as if in thick cloud.

"Dougie can you hear me?"

"But... but..."

The alternator had failed as the engine wound down, but the Gnat (unlike my F4 Phantom incident) had a battery as back-up. At least I could speak to people, and the flying instruments worked.

"Dougie get a grip! We've had an engine failure."

"But...but..."

"*Mayday...Mayday...Mayday,*" I called on the radio's emergency frequency. "*Gnat 20...single-engine aircraft...engine failure 15 miles north-west of Llanbedr. Descending through 26,000 feet. Request forced landing at Llanbedr.*"

"Dougie can you hear me?" I asked again, after transmitting the '*Mayday*' call.

"But...but" Dougie still seemed shocked and overwhelmed.

"Gnat 20 this is the Emergency Controller. Your Mayday acknowledged. Standby."

After the Emergency Controller had spoken, I continued my conversation with Dougie: "I've done some of the flip-card emergency drills already; I'm about to do the rest. Head 120 for Llanbedr. Maintain this speed."

"OK...OK." Dougie appeared to be emerging from his state of immobilisation.

"It'll be alright Dougie. Just keep flying while I do the drills." I carried out the remainder of the emergency procedures from memory, then checked the emergency 'flip cards' to ensure that nothing had been missed. The altimeter showed us to be descending rapidly.

"Gnat 20 this is the Emergency Controller. I'm speaking to Llanbedr now. Continue on your present heading. Standby."

We were approaching 15,000 feet. Half of our initial height had been lost already.

"Dougie, we're going to try to re-light the engine," I said at that point. The altimeter continued its remorseless wind-down. We were still iced-up, and could not see out of the cockpit properly.

"I'll do the re-light drills, Dougie. You just keep flying towards Llanbedr." I read out the in-flight start checks. At the end of the check list, the drill called for the starter button to be pressed.

I pressed the starter button.

Without hesitation, the engine re-started. The cockpit heating resumed, and the iced-up canopy began to clear. I could see Llanbedr airfield appearing on the nose, and to our left was the distinctive outline of the Snowdonia high ground; to our right was the sea.

"What happened?" asked Dougie.

"I think we'll be OK now," I said. "It was the engine objecting to the slam check, I guess. I suggest we return to Valley for a practice flame-out pattern. We'll then be 'in the groove' if the engine flames-out again."

With the engine working again, and the canopy clear, I called Air Traffic Control: "Gnat 20. Our engine has re-lit. Down-grading from Mayday to Pan. Request priority flame-out pattern at Valley." The code-word 'Pan' meant that we had down-graded our emergency status.

"Are you happy to carry on with the flying?" I asked Dougie.

"Yeah. I'm OK. I'm OK," said Dougie.

We caught some excited chatter on the radio, as other Gnats had picked up our emergency call. An Instructor flying in the area homed-in on us, and acted as escort for our return to Valley. As we approached Valley, Dougie maintained a higher altitude than normal, and I talked him into a good position for a 'flame-out' pattern. Once in the pattern, we felt relieved. We could have landed safely, even if there was a further problem with the engine.

As it turned out, the engine gave no more trouble, and we landed unscathed.

After the landing, and during the inescapable paper-work and form-filling following the incident, I was quietly aware that my sinuses had objected to the fast descent. Fortunately there had not been any nose-bleeding.

I managed to continue with a normal flying routine for two months.

However, in the Spring of 1977, I had further sinus problems, which resulted in nose-bleeding. The Doctor decided to give the sinuses a complete break. "The air-pressure changes caused by such regular flying never give the sinuses a chance to settle down," he said. I was grounded for 3 months.

During the 3 months, I was employed as a Simulator Instructor. It was a difficult period as the possibility of never flying again played on my mind. The uncertainty of the situation exacerbated the sense of impending doom.

On the last day of May, I returned to flying duties. It lasted for 3 weeks. My final fixed-wing flight as aircrew was flown on the 20th of June, 1977. At the end of the flight, I had no option but to see the Doctor again.

"That's the end of the road, I'm afraid," said the Doctor. "I've no choice. As of now, your medical category is withdrawn. You are permanently grounded subject to a review by a Medical Board."

The Medical Board, when it happened, turned out to be an Air Commodore at the Central Medical Establishment in London. The Air Commodore was a rotund gentleman who looked at me gloomily over his half-moon glasses.

"Do you want to carry on a flying career?" he asked.

I assured him that I did.

"Further fast-jet flying is out of the question now. But I suppose there are a few options which don't involve rapid descents. How do you fancy VC10s?"

My face must have dropped.

"Hum!" he said grumpily. "Can't be too fussy you know. But I suppose it would be a bit dull after what you've been doing."

The Air Commodore gave me a piercing look.

"We could maybe try to wangle helicopters, although the finance boys won't like it. From the medical point of view, though, it should be a good solution as its a purely low-level role." The Air Commodore noted that I appeared to look rather more interested.

"OK. Leave it to me," he said. "I'll see what we can do."

I left the building of the Central Medical Establishment with rather mixed feelings.

When I got back to Valley, I discussed it with colleagues. "That's great," they said. "They're a weird lot in the helicopter world. Should suit you fine."

"Thanks," I said.

"At least you'll keep flying, and fun flying at that. And there's the North Sea Oil just starting. You'll get a good job there, and they pay those boys a fortune."

Half of me rued the end of fast-jet flying, which had seen so many remarkable experiences. The other half of me looked to the future. When the posting notice came through some weeks after the visit to the Central Medical Establishment, my prospective life was confirmed. I was to join number 181 Helicopter Conversion Course at Shawbury in Shropshire.

I was destined to become a helicopter pilot.

CHAPTER 26

Whirlwind Encounter

"The idea of this sortie is to teach you the art of hovering," said Carl, my helicopter Instructor at Shawbury. He had a reddish complexion, and a look of permanent amusement. An experienced helicopter man, he had an easy-going approach with his students.

It was Autumn 1977. We were about to fly a Whirlwind helicopter, registration XR458.

"What we'll do," my Instructor had said in the pre-flight briefing, "is go over to the far side of the airfield, and try to find a quiet corner. We'll spend about 45 minutes there, so you can practise hovering. We'll also look at other aspects of flight in the hover."

In the dispersal area, I ran through the pre-start checks, and released the rotor-brake when the engine had started. Above us the main rotor began to turn. The noise and vibration progressively increased. The engine instruments reacted. The torque-meter reading increased. The tail-rotor started to rotate.

Already, I had flown some familiarisation sorties. I had found them quite eccentric. The flying was generally more relaxed than fast-jet, but there were many hidden hazards.

"I have control," said Carl, and he ground-taxied the Whirlwind past other helicopters in the dispersal area. He then lifted the helicopter into the hover, and obtained clearance from Air Traffic Control to move to the far side of the airfield.

As I took over the flying controls from Carl, he said: "Relax: you're too tensed up." The Whirlwind was unsteady as I attempted to keep it in the hover. "It's quite twitchy," said Carl. "Try not to over-control."

I held the cyclic stick lightly, and made continuous small adjustments to hold the hover in a steady position. As soon as my hand-grip tightened, it was a sign of becoming too tensed-up; hovering became harder.

As well as holding the cyclic stick in my right hand (the equivalent of a Fixed-Wing aircraft's control column), in my left hand was the collective lever. I moved it vertically to adjust our height up or down: the rotors were moved 'collectively'. By my feet, the rudder pedals had to be moved in synchronisation with the other two controls; that regulated the tendency to 'yaw'.

"It's just a bit tricky until you get used to it," said Carl.

"Imagine an inverted saucer," continued Carl. "Then assume you are a block of ice trying to stay on top of the saucer. As soon as the saucer moves, you have to react before you slip off." He looked across at me. "Holding a hover is the same general idea," he said.

"Cool," I said.

"Very," he replied.

Hidden away in a corner of the airfield, it felt bizarre that I was legitimately spending flying time going nowhere. My mind had became drilled to the idea that when airborne, it was necessary to move from 'A' to 'B' rapidly, preferably very rapidly.

We had chosen a grassy area, clear of trees, and the Autumn sun shone through the Whirlwind's canopy. Frightened rabbits occasionally popped-up their heads to look at us before they raced off; the scene was far removed from the fast-jet world.

"Fun, isn't it?" commented Carl.

"Are we really logging this as flying time?" I asked him.

"Too right," he replied. He glanced at me. "But don't get carried away," he grinned. "It gets more difficult soon."

We then tried moving the Whirlwind sideways. Height control became unsteady, and the co-ordination of the rudder pedals was a problem too.

"It'll become second nature quite quickly," said Carl. "But for now you'll have to remember to steady-down the control inputs. That last side-ways movement was OK, except that you were slow reacting with the rudder pedals. The tail-rotor is very powerful. That's why the nose started to swing around."

"There are too many things to do all at once," I moaned.

"You ex-plank pilots don't know the half of it," said Carl.

Nearby I spotted a course colleague practising the same exercise.

"Don't forget the 'look-out'," said Carl. "Before moving anywhere, you should turn on the spot, and check the area is clear all around." I continued to hold a steady hover. "Which brings us neatly on to the next thing," said Carl, "spot turns."

Carl took the controls, and demonstrated a perfect turn through 360 degrees on the same spot. The rate of turn was at an even pace, our height remained constant, and we did not move an inch from the chosen spot.

"Now you have a go," said Carl.

Initially, my spot turn went well. Half-way round the turn, however, when we were down-wind, the rate of turn suddenly started to speed up. I over-reacted with the rudder pedals which made matters worse. The Whirlwind began to wobble and wander. Carl took control again, and put the machine back into a steady hover.

"You have to watch that," said Carl. "The down-wind effect can be dramatic." He then handed the controls back to me for more 'spot turn' practise.

"OK. We'll move on to the next exercise," said Carl eventually. "Which is to fly backwards."

"First of all," said Carl, "check the 'six-o'clock' area is clear." I applied rudder pedal to turn the Whirlwind through 90 degrees. "That looks clear," said Carl, "now turn back on to our original heading." I applied rudder pedal in the opposite direction.

"Good," said Carl. "Now increase our height slightly, and then with a small backwards input on the cyclic, start the machine moving." I kept the Whirlwind on a steady heading with the rudder pedals, and gingerly applied the cyclic stick backwards. Slowly, the machine started its rearwards progression; it felt unreal. For an ex-fast jet pilot, the sensation seemed almost mind blowing. My head moved constantly as I looked about me, struggling to avoid a sense of disorientation.

"It's important to keep the speed down when you fly backwards," said Carl. "Otherwise you can end up out of control."

After moving the Whirlwind backwards for a few yards, Carl told me to return to the hover. He sensed that I had found the reverse-gear bit rather unnerving, and told me to hold a steady hover for a few moments.

"It's an odd sensation at first," said Carl at length. "Don't forget the look-out going backwards. Use the mirrors, and use the 'crewman' if there is one." Carl checked his watch. "Anyway, time's-up for now," he said. "You can take me back to dispersal."

After clearance from Air Traffic Control, I moved the helicopter with some apprehension towards the busy dispersal area. "OK. We'll land here, just before the dispersal area," said Carl, "and ground-taxi the last bit to prevent our down-wash from disturbing other machines."

We had a detailed de-brief after landing, and Carl then summed-up: "Well done," he said. "That went pretty well on the whole." He raised his eyebrows, and smiled. "After the week-end, we'll move on to the next exercise: 'EOLs'."

"EOLs?" I asked.

"Engine Off Landings," replied Carl.

"But we've only got one engine," I pointed out.

"You'd better not screw it up then!" Carl was still smiling.

As I left the training set-up to walk home to our Married Quarter, the noise from a helicopter pierced through the surrounding environment; the machine was being ground run for engineering tests. Aircrew and groundcrew were regularly warned about the potential damage to their hearing; ear protectors should be worn at all times. Gradually the sound diminished as I approached the 'Married Patch'.

A noisy scene, however, greeted me when I got home. Sue was feeding the children. Giggles and squeals filled the air in our Married Quarter at Shawbury.

"Hey-o Daddy," called Lizzie. Nearly three, she was also the spokesperson for her 18-month old brother. "What you bin doing?"

"Flying helicopters," I replied.

"Why?" she asked. I looked at Sue; she shrugged.

"It's fun," I replied. Lizzie looked at me doubtfully.

"Why?" she asked again.

"Maybe we'll go on a picnic tomorrow," I said, avoiding her question. "It's Saturday, and the weather will be nice."

"Yeaaah…" bellowed Lizzie, a chorus quickly taken up by her brother, though he was uncertain why.

As we drove towards Oswestry on that Saturday morning, the mountains of the Cambrians showed in the distance. Their prominent outline dominated the horizon, a graphic and stark image. Trees covered the high ground, and the Autumn colours were vivid; there was a timeless feeling.

After crossing the Welsh border, we saw a road sign to Llanhaeadrym-Mochnant. Passing through the Welsh villages, we eventually drove the car along winding roads to Lake Bala. Set deep in a valley, the pencilled-shaped Lake was around four miles long. It had been a popular feature for navigation exercises when I had flown Gnats from Valley.

"This looks great," said Sue. We parked, and unloaded by the side of the Lake. Lizzie and Alan raced around in excitement. Their mother and I prepared the picnic. Sue handed me some unappetising items of salad to sort out. A lettuce had been squashed; a casualty of the journey.

"What's on the agenda for next week?" Sue asked eventually.

"EOLs" I replied.

"End of Lettuce?" she enquired.

"Extremely Odd Lifestyle," I said.

"You're not wrong there!" laughed Sue.

CHAPTER 27

Called to the Rescue

"That's it!" said Carl. "We're now committed." He had brought back the 'Speed Select Lever' of our Whirlwind helicopter to the 'Ground Idle' position. The engine was no longer driving the rotor blades. As we descended, the upward flow of air through the blades caused the rotor blades to keep turning.

"In auto-rotation," the Ground School Instructor had told us, "the rotor blades are turned in the same way as sycamore 'wings'." As a visual aid, the Instructor had produced twin-bladed sycamore wings, held them high above his head, and then released them. The wings rotated at an even pace as they descended to the floor.

"Unlike the poor sycamore wings," said the Instructor, "in a helicopter, just before landing from the auto-rotation, the pilot raises his collective lever to cushion the touch-down. The rotors will immediately start to slow down when the collective lever is raised. The timing is therefore crucial." The course students looked at each other anxiously.

The Ground School Instructor then continued: "When the helicopter is in auto-rotation, the height of the machine (Potential Energy) is traded for the rotor head's Kinetic Energy." The students nodded their heads judiciously.

As Carl and I put the theory into practise, I noticed that he was a little more up-tight than normal. We were practising our 'Engine Off Landings' at Turnhouse, an airfield near Shawbury.

"Watch your speed!" Carl said to me as the Whirlwind fluttered downwards. "And concentrate on that level surface we've been aiming for."

"Passing 1,000 feet," called Carl. I was monitoring the rotor revolutions during the descent. Occasionally they went too high, and an ominous high-pitched whine from the gear-box could be heard. A small upward movement of the collective lever quickly brought the rotor revolutions within limits.

"Approaching 500 feet," said Carl. I aimed to keep the airspeed steady, and used small tail-rotor movements for directional control.

"Approaching 300 feet," Carl called out. "Standby!" I tried to force myself not to grip the controls too tightly.

"250 feet," said Carl. "Start your flare…*now*." I raised the nose of the Whirlwind by pulling back on the cyclic stick.

"That's far enough," said Carl. The rate of descent had been slowed. "Now level the aircraft." I pushed the cyclic stick forward again.

"200 feet," called Carl. "Standby with the collective."

The ground appeared to be racing towards us.

"*Standby…pull the collective lever… now!*" said Carl. As I raised the collective lever to its full extent, the effect was dramatic. The rate of descent was suddenly checked. We had the illusion of almost entering the hover. It was the final use of the Kinetic Energy stored in the rotor head; it controlled our touch-down.

The Whirlwind landed on the grass with a thump. "Don't pull the cyclic stick back now," said Carl. "That's an old fixed-wing habit you're going to have to lose." As the Whirlwind continued to roll along the grass, I gently applied the wheel brakes.

"Just bring the machine to a halt here," said Carl. "And I'll restore the engine." Carl then slowly advanced the 'Speed Select Lever', the device which controlled inputs to the engine to keep engine revolutions within a defined band.

"Not bad," said Carl. "A little firm on touch-down perhaps. Let's take-off again, and try another one."

Once the 'Speed Select Lever' had been returned to the 'Flight' position, I lifted the Whirlwind into the hover, and spot-turned through 90 degrees.

"All clear behind," said Carl.

We called Air Traffic Control at Turnhouse for take-off clearance, and I pushed the cyclic stick forward to allow the airspeed to build up during the take-off.

Suddenly, half-way through the take-off procedure, Air Traffic Control called us: "Whirlwind Tango, this is Turnhouse."

"Go ahead," said Carl.

"Sorry to interrupt your take-off," said Air Traffic Control, "but I have an urgent Police message." Carl looked at me with a puzzled expression.

"Ready to copy," said Carl to Air Traffic Control.

"Roger", said the Controller. "The Police have received reports of a road accident just west of here. The telephone system's been giving problems, and they're having difficulty with the exact location. Can you assist?"

"No bother," said Carl, as he took the flying controls. He applied a high angle of bank as he turned on to a westerly heading.

"Check the fuel, please," Carl said to me. "And work out an available 'time on task'."

As I was making the calculations, Carl maintained a height of 500 feet, and flew at 90 knots airspeed. "When we get further from the Airfield," he said, "we'll slow down to around 50 knots which should be a good search speed."

I had worked out that we had 45 minutes available before we needed to return for refuelling.

"That's fine," Carl said to me. "Thanks."

As we reduced airspeed to 50 knots, we searched all around. There were tall trees which restricted our view. The ground was covered in brightly coloured autumn leaves.

"Let's head north up this road," Carl said eventually. "The area looks quite tree-covered; it could make the roads slippery." At one time, Carl had been a Search and Rescue pilot; his experience meant he had developed a sixth sense.

We continued to search on either side of the Whirlwind. However, we still saw no signs of the Traffic Accident.

Suddenly, Carl said: "What's happened over there?"

To our right, there was a narrow lane. It was twisting, and covered in leaves. Carl pointed the Whirlwind towards the lane, and began to reduce our height. I then saw what had caught his attention. Just by some trees, a small group of people were gathered. As we approached them, someone waved. Then we noticed a car in the ditch beside the leaf-covered lane. The car was on its roof.

"Work out our 'Georef' position, please," said Carl. I checked the 50,000 scale map, and wrote down our position.

"OK," said Carl. "We'll climb back up to 500 feet so I can talk to Turnhouse. We'll then land in that adjacent field, and I'd like you to get out and speak to those people."

Once Carl had passed our position to Air Traffic Control, he entered a descending orbit around the field he had chosen.

"Run through the pre-landing checks, please," he asked me.

I ran through the checks, and Carl began an into-wind approach towards the field.

"Check your side for wires and other obstructions," called Carl.

"It looks clear," I replied.

"Fine," he said. "I'll land on that flat area just ahead."

"It still looks clear my side," I said to Carl. He brought the Whirlwind to a hover over his chosen site, and lowered the aircraft slowly on to the grass.

"It's not too bad," I said. "But the wheels are sinking a little."

"Roger," said Carl. "I'll have to keep 'turning and burning'." Carl needed to keep the rotors turning; then he would be able to control any tendency of the helicopter to sink further into the mud. "You can nip out now," said Carl.

The group of people we had seen from the air, had watched the helicopter make its approach. Two people came towards me as I climbed out of the Whirlwind's cockpit. Above the noise of the helicopter I asked if there was anyone injured.

"The driver's trapped," they said. "She's a young lass, and is in quite a bit of pain."

As I walked to the car, I noticed a smell of petrol. "Have you a fire extinguisher handy?" I asked.

"No," they replied.

I ran back to the Whirlwind, and hastily told Carl what was happening.

"We haven't any 'Entonox' gas for pain relief," said Carl, applying his old Search and Rescue skills. "I'd be reluctant to give her tablets from the First Aid Kit in case she needs surgery. I reckon you should use our foam fire extinguisher right away, though."

I released the catch securing the fire extinguisher to the helicopter's bulkhead, and dashed back to the upturned car. The trapped driver was being comforted by a middle-aged lady. "Cover your faces," I told them. I then pressed the catch on the side of the fire extinguisher. White foam erupted from the extinguisher. In the distance, I thought I caught the sound of wailing sirens above the noise of the helicopter.

I ran back to the helicopter. "The emergency vehicles seem to be in the area," I said to Carl. "I thought I heard their sirens. If I fire a couple of emergency flares, it should help them to pinpoint our position."

"Yeah. Good plan," said Carl.

I unclipped the revolver for the emergency flares, and took two of the bulky flares to the centre of the field. I loaded one of the flares into the revolver. Then I aimed the revolver into the air, and squeezed the trigger. A bright red firework shot skywards, leaving a trail of coloured smoke. The second soon followed.

After a few minutes, from around the corner of the lane, we saw flashing blue lights on top of the emergency vehicles.

"You'd better speak to them," said Carl.

I hurried back to the scene of the accident. A paramedic was jumping out of his ambulance. "Saw your flares!" he said to me.

The paramedic ran to the trapped driver, and spoke to her. Firemen quickly unloaded their cutting gear. Even above the noise of the helicopter, the cutting equipment could be heard as the firemen efficiently released the driver.

"Your chopper's a training machine, isn't it?" the paramedic asked me.

"Yes," I replied, "And I'm afraid we haven't stretchers or anything."

"OK," said the paramedic. "In that case we'll take her to hospital in the ambulance. There's a problem with her back."

The paramedics had placed a neck collar around the patient. The young woman was strapped into a stretcher. She half-raised one hand in a feeble wave, and her pale face attempted a smile.

As they loaded the casualty into the ambulance, I went back to the Whirlwind and relayed the latest news to Carl.

"I can see the ambulance leaving now," he said. "We may as well return to base."

After a pause, Carl said: "Anyway, I can't stand any more excitement for today!"

Carl then offered to hand control of the Whirlwind over to me. He said: "There's a clear area just ahead for take-off, and we're still pointing into the wind. Are you happy to do the flying?"

"Sure am," I confirmed.

Carl read-out the before take-off checks. I then lifted the helicopter into the hover, checked the area behind, and Carl made a 'blind call' on the radio to inform any passing aircraft of our intention. After the take-off from the field, I levelled the aircraft at 1000 feet for our transit flight back to Shawbury.

As we passed Turnhouse, Carl suddenly looked at me and said: "I suppose we could just squeeze in another practice EOL." He grinned.

Carl glanced at me again: "Only joking," he said. Carl was still grinning.

CHAPTER 28

In the Dunker

"Before you start training in Search and Rescue techniques," the Briefing Officer at Shawbury had said, "You'll be required to do the dunker."

Run by the Royal Navy at Portsmouth, the 'dunker' was an eccentric-looking contraption hanging above a swimming pool. The device was roughly the size and shape of a helicopter cockpit. With students strapped in to seats placed inside the cockpit mock-up, the dunker was lowered slowly into the swimming pool by an over-head crane. Once in the water, the dunker was designed to roll-over on its side. The students then had the opportunity to practise the safest method of escape. From the start of the dunker's journey, to the students escaping, took just a few minutes.

"The best approach," the Royal Navy Instructor had said, "is to take a deep breath when the water reaches neck level, and then deliberately put your head below water. It can get kind of unnerving if you leave it to the last second before submerging your head."

We were not allowed to un-strap, or attempt to escape, until the dunker was felt to bump against the bottom of the swimming pool. "At that point," said the Instructor, "you can safely assume that the helicopter rotors have been stopped by the rising water. If you try to escape too early, there is the danger of decapitation."

For our Course, the Instructor had persuaded his wife to join us; he was keen to show her that the procedures were safe. His wife was outwardly cheerful, but forced laughter revealed her true feelings.

"OK Ladies and Gents," the Instructor had said on the day, "Are you all ready?" The students, duly seated and strapped-in, gave 'thumbs-up' signs.

"Take it away, Bill," said the Instructor. The crane operator pushed a switch to start the dunker's passage into the pool.

Standing inside the dunker cabin was a diver, just a few feet from the students' faces. When under water, the poor visibility meant that the students would become oblivious of his presence. Another diver was already in the swimming pool.

The whine of the crane's electric motor shrilled through the pool as the dunker was lowered progressively towards the water. A flippant remark caused a nervous snigger amongst the students; they no doubt sensed the inevitability of their immediate fate with some angst. The diver shifted his position: "Standby everyone," he said. "Just remember your training." The diver lowered goggles over his eyes.

The water was felt by our feet as the dunker continued its downward journey. Next to me sat two other students. They had to escape before my path was clear. The prospect added to my feeling of tension.

As the water level began to creep up our legs, I looked around briefly at the other students. Their expressions revealed focused minds. The Instructor's wife appeared to be nervous. Her face was white and uneasy.

The water felt colder as we became further immersed. We wore orange-coloured boiler suits which offered some protection against the cold water, but still I shivered as the water level changed. I had an 'orientation arm' out-stretched.

"You may find this hard to believe sitting here in a warm classroom," the Instructor had said. "But once you're fully submerged, within seconds you'll become dis-orientated. You won't know if you're on your arse or your elbow." The solution therefore was to hold on to a part of the helicopter fuselage. The 'orientation arm' pointed to the escape route. When it was time to unstrap and abandon the machine, the direction of escape was shown by the 'orientation arm'.

The high-pitched whine of the crane's electric motor became slightly louder. Perhaps the motor took more strain as the dunker moved under the water's surface. The student next to me felt for the position of his seat-buckle, by then submerged.

The water level rose towards the upper part of our chests. My immediate neighbour had used his left arm for 'orientation'; his escape route was away from my position. The student next to him had to move before my neighbour could escape. I was conscious again of my place, last in the pecking order. I also had to use my left arm for 'orientation'; somehow it felt wrong.

The echoes of noises around the swimming pool became more pronounced as our heads approached the pool's surface. The over-riding sound was still the whine of the crane's electric motor, but other minor clamours were still heard. At the far end of the pool, a member of staff dropped something. A peculiar 'clank' bounced around the swimming pool.

I thought there was a faint cry from one of the students as the water level approached our lower necks. Quickly, I looked across at the Instructor's wife. She was struggling with an on-rush of last minute panic. Her face was screwed up; I imagined her to be suffering a claustrophobic nightmare.

The water level by then was lapping against the under-side of our chins. There was nothing to be done to help the poor woman. Any second now, each person would be faced with an overwhelming individual force. My neighbour took a deep breath; his decision had been taken. He plunged his face with some violence under the surface of the water.

Slightly taller than my neighbour, I decided to wait a few more seconds. As the last in line, I had to hold my breath for longer than the two students next to me. In any case, I was unhappy under water; it was not a natural feeling environment for me. Furthermore, at the back of my mind was a concern that the water pressures would cause a resurgence of sinus problems. I felt the best tactic in my case was patience before taking the plunge.

Suddenly, I was startled to find some water entering my mouth. The taste of the chlorinated water was sour. I had one hand over my seat-buckle; my 'orientation' arm was fully extended. I longed for a third arm to wipe away the foul-tasting water from my mouth.

At that moment, I had an impulsive and crazy notion. I yearned to be something different. I seemed to enter a sub-world, away from normal levels of consciousness. Gradually, my mind began to lose its grip of reality. I gave the life around the pool a cursory farewell; I felt the need to concentrate on my new world.

The light became brighter. It was a strange, unnatural light. There were signs of colour within the light, but they were subtle and muted. The colours kept changing, they almost appeared to take up shapes.

In the middle distance, I noticed there was someone sitting. I felt an urge to join the person, but found it hard to move. My arms felt bound.

There were sounds, but they were unobtrusive. The main sensation was the unique light. It had a hypnotic effect.

As I struggled to free my arms, I thought that the person sitting appeared to look in my direction. I tried to speak, but it was impossible. In spite of the feeling of helplessness, I was relaxed. There seemed to be no hurry to do anything; I was at ease.

Then I noticed that the person had stood up. Still looking in my direction, the person had started to move towards me, slowly and deliberately. I had no fear; the person was friendly according to my sub-conscious mind.

Still the peculiar light danced around me as the person took measured steps in my direction. I was keen to join the person, to show friendship, but despite my efforts, I was unable to move. Some of the previously subdued sounds became slightly more pronounced. There was no speech, however; just background noise.

The person appeared to lift an arm slowly. Continuing to walk towards me, the outstretched arm was welcoming. The hand almost seemed to beckon. Progressively, the floating person became surrounded by a different light. I felt convinced the person was trying to reach me, but progress was somehow impeded. Again, I struggled to free my arms.

The shape of the person appeared to change. Still with outstretched arm, the individual seemed to have further difficulty in reaching me. I could not distinguish a face, but the colours surrounding the person became more spectacular. The outstretched arm moved sideways; the person was attempting to swim. The progress towards me had become painfully slow. Defying a sense of frustration, I tried to reach out to the person.

I still felt a deep tranquillity. The person's outline became blurred as the attempts to reach me became more laboured. *Even the fantastic light had weakened.*

Suddenly, I felt a sharp knock on my 'orientation arm'. I could not see him, but I was convinced that the knock on my arm was from the diver. The sub-world still dominated my mind; the time seemed to have stretched to infinity, but in reality it had been just a few seconds.

I had left my last intake of breath a little too late. Nevertheless, I had submerged my head just in time, and so far had managed to control my breath successfully under water. As feared, however, I was aware of sinus pains.

The bump of the dunker against the bottom of the swimming pool had been felt. My 'orientation arm' had been close to my neighbour, and I was conscious of his efforts to escape. However, I could not see my neighbour, and I was reluctant to move until I knew for certain that he had gone. I had delayed too long. I had sat still; I had become mesmerised by the situation.

There was another knock on my arm, sharper than before. It brought me round to reality; it was the prompt I needed. My confused mind suddenly became determined to follow the rigorous training procedures. It was as if I had had my auto-pilot rapidly switched on. My right hand twisted the seat buckle. The hand groped hurriedly with the seat straps, to move them away. I then lifted myself clear of the seat with my right hand. I followed the direction of my 'orientation arm'; the arm was still resolutely holding on to part of the dunker.

My right hand made swimming motions towards the escape route. I confirmed that the space next to me was clear, and soon I was able to release my 'orientation arm'. Both hands then swam energetically in the direction indicated by the 'orientation arm'.

In a dramatic-feeling splash, I suddenly broke through the surface of the water. My eyes took a few moments to adjust; they felt painful and stinging. It was hard to take-in the scene around me. My sinuses were sore; I checked for signs of bleeding, but thankfully found none. The sounds of the swimming pool were loud and intrusive.

On the side of the pool I noticed that the Instructor's wife was being comforted by her husband. She was slumped forward, and appeared to be in tears.

One of the divers was swimming near me. "You alright mate?" he asked.

I spluttered a reply.

"You were last out," continued the diver. "I thought you were stuck."

Still I was unable to speak properly.

"Dodgy business, this," said the diver. "It's amazing what people get up to down there."

CHAPTER 29

Learning from Lundy

"Watch out!" said Ralph. I moved the Whirlwind's cyclic stick slightly.

"Not that one," said Ralph. "We're getting too low over the water." I immediately applied an upward movement to the helicopter's collective lever. I gripped the lever tightly in my left hand.

"You're too tensed up," said Ralph. "Try to settle down."

We were practising hover techniques over water. Without the solid visual references used during a land based hover, it was proving to be difficult.

"Look at the far horizon," said Ralph. "That'll give you a general feel. Then search for something on the water's surface: a bit of seaweed, some spume, waves perhaps. Then try to use all these visual references in combination. Also make use of the aircraft's radio altimeter. It's very accurate and will tell you when you're sinking too low. The 'radalt' is especially useful in murky conditions."

Ralph was an Instructor with the Search and Rescue Training School, based at Valley in Anglesey. It was the early part of 1978, and along with my fellow students from Shawbury, I had been detached to Valley for two weeks to learn the fundamentals of Search and Rescue flying. Ralph had many years of experience in Search and Rescue work. He was slightly balding; his impassive approach was no-nonsense, but nevertheless friendly.

"That's better," said Ralph. "What we'll do in a minute is throw out a 'drum', and you can practise winching it on board."

In the rear of the Whirlwind's cabin we had a crewman. He was on a specially-designed strap which allowed him a certain freedom of movement. However, if he was knocked, or if he slipped, the strap would hold him in the helicopter.

"OK Bob," Ralph said to our crewman. "You can despatch the drum now."

Bob checked beneath the helicopter, then kicked the battered oil-drum out of the cabin. The drum was painted in bright 'dayglow' colours.

Bob checked that the airspace around the helicopter was clear, then he called: "Clear above and behind."

It was the signal for me to start a take-off in to the wind. I pushed the Whirlwind's cyclic stick slightly forward, and as the nose dipped down, the aircraft began to pick up airspeed.

"Remember to level at exactly 100 feet," said Ralph. "Accurate flying is most important." When the radio altimeter approached 100 feet, a co-ordination of collective and cyclic control inputs checked our height.

"We'll make this a left-hand circuit," said Ralph. "So when you're ready, apply 30 degrees angle of bank."

I looked over my shoulder to ensure the area was clear, then moved the cyclic stick to the left.

"Use the instruments as well as looking out," said Ralph. "It's easy to become dis-orientated, especially in foggy conditions."

As we turned downwind, all three crew members searched for the drum. There was a swell on the surface of the sea. The drum quickly became hard to spot.

"Downwind," I called.

"Roger," said the crewman. "Still looking for the drum."

I asked Ralph to read out the downwind checks. He rapidly rattled them off; *"Hoist Master to 'Crew'…temperatures and pressures… fuel… harnesses."* Bob called: "Still searching."

"Roger," I replied. "Continuing downwind."

"Estimate target now in the left nine o'clock," said Bob. The surface swell made it hard for the crew to see the drum, but we caught occasional glimpses.

"Estimate target now coming to the eight o'clock," said Bob.

"Roger. Turning in," I replied.

During the 'turn in' I asked Ralph to give me the 'Finals checks'.

"Visual with the target, now coming to the eleven o'clock," called Bob. "Range eighty."

"The target ranges given by the crewman," the Briefing Officer at Valley had said, "are not related to a specific scale. They are purely an indication, in the Crewman's judgement, of the relative distance between you and the target."

"Target range seventy. Safe height…no lower" said Bob.

"Target sighted," I said.

In spite of my call, the 'dayglow' coloured oil drum was still hidden by the sea's swell from time to time. The crewman's main concern was to watch the drum. The two pilots, however, had to monitor the aircraft instruments; sometimes they lost sight of the drum. The pilots concentrated on gradually reducing airspeed and height as they approached the target. They had to find the right balance: too fast, and they would fly past the drum; too slow and they wasted time reaching a potentially desperate survivor.

"Think of the 'Lundy Island' incident," Ralph had said to me sometime later. "Every last second counted then."

❖ ❖ ❖

"Coastguard… Coastguard… this is Lundy Island. Do you read? Do you read? Over." The Coastguard Duty Officer immediately sensed the urgent tone of the radio call.

"Lundy… Lundy… you are weak. Go ahead. I say again, go ahead. Over." The radio crackled intermittently as the Duty Officer strained to listen to the message.

"Coastguard… Coastguard… we have an emergency… I say again, we have an emergency. Over."

"Lundy… Lundy… go ahead…go ahead. Over."

"Roger, Coastguard. Seriously ill female requires evacuation. I say again evacuation. Did you read. Over?" The Duty Officer wrote some details on his note-pad.

"Lundy… Lundy… message copied. Standby. Over."

The Duty Coastguard Officer called across to his colleague: "Fred!"

The colleague looked up. "There's a problem at Lundy, Fred. Female requires urgent evacuation."

"OK," replied the colleague. He checked the time on the Operations Room clock. "I'll scramble the Chivenor Whirlwind," he said.

The Whirlwind helicopter from Chivenor had been duly 'scrambled'. It had flown to Lundy Island, located the seriously ill woman, and had evacuated her from the Island. There had been hassles, as is often the case with such flights, but the casualty had been rescued, and was being flown to safety.

However, during the return flight from Lundy to the mainland, with the casualty on board, disaster had struck. The Whirlwind had suffered engine failure. The crew were forced to ditch the aircraft into the sea. The rescuers had themselves become casualties.

A second helicopter from Chivenor had been 'scrambled' immediately. The relief Whirlwind had spotted the survivors shortly after departing from Chivenor. Fortunately, the visibility that day had been good. The Captain of the helicopter was under no illusions; his speedy arrival was a matter of life and death. He flew the aircraft low over the sea. He flew at maximum speed, maximum power, maximum everything.

It must have been agonising for the rescue crew. In the mid-distance, they could clearly see the casualties. The slow progress of seconds and minutes had no doubt tormented them.

By the time the 'rescuers of the rescuers' had reached the scene, the survivors were in a desperate situation. The female casualty had nearly died. Apart from her original problems, she had become hypothermic, and heavily traumatised by the whole incident. It was thanks to the heroic efforts of the crew, her fellow casualties, that she had been kept afloat and alive.

Eventually, when the second Whirlwind had reached the scene, it was in an unholy rush. At that stage, the Captain had no doubt forgotten some of the niceties of a neat training scenario. He had practically hurled the aircraft into position. In double-quick time, the Winch Operator had the Crewman winched down.

The crews of both the Whirlwinds had been highly professional in their duties. The woman victim was first to be winched aboard, soon followed by the Crewman of the first helicopter. That Crewman had continued then to administer first-aid to the female casualty, while the second crew concentrated on rescuing the remaining survivors.

It had been touch and go, but the life of the original casualty was saved, along with the lives of all the first aircraft's crew.

❖ ❖ ❖

"Target now on the nose, range fifty," called Bob.

"Relax!" Ralph looked across the cockpit at me. "Don't get too tensed up." I had to keep reminding myself to relax. I was nervous; we would shortly enter another hover over the sea.

"Target still on the nose, now range forty," said our Crewman. "Winching out." He operated the winch switch. At the end of the winch wire was a hook; Bob aimed to lower the hook to a point just above the wave tops.

I glanced back at him. Bob was kneeling in the open door-way at the front of the Whirlwind. He was securely held by the crewman's harness. On his head, Bob had a bright yellow 'bone dome'. From the top part of the 'bone dome' a sheet of clear perspex protected his face from the effect of the biting wind. Bob wore a special heavy-duty flying suit, and a life jacket on top of his flying suit. He had one hand operating the winch switch. His other hand was above his head, gripping a handle attached to the helicopter's bulkhead. When we were in the hover, this hand would be used to guide the winch wire; for that he wore a specially reinforced glove. As first on the scene in many unpleasant situations, the crewmen were highly trained and courageous individuals.

"Target moving slightly left, at a range of thirty," said Bob.

Ralph and I could see the drum more clearly now. I made a small correction with the cyclic stick. It brought the relative position of the drum back on to our nose.

"Target now twelve o'clock," called Bob.

In the middle distance, we could just make out the coast-line of Anglesey. The visibility was fair, but we lacked a clear horizon. On the aircraft radio, we listened to the flying in progress at Valley.

"Target still on the nose, at a range of twenty," said Bob.

Ralph looked across at me. "From now on," he said, "you'll find the target can move fairly quickly away from our centre-line."

As if on cue, Bob called: "Target now moving right of the nose as we approach range fifteen."

The drum was bobbing around significantly in the sea swell, and I had to avoid a temptation to over-control. I was still gradually reducing the helicopter's forward speed.

"Well done," said Ralph. "That's a good steady approach speed. The thing to beware of now is over-doing the speed reduction. If we enter the hover too early, it can be awkward."

"Target range ten," said Bob. "Still slightly right of the nose."

The movement of the drum appeared to be quite violent in the sea swell. It became more noticeable as our approach took us closer.

"Try to correct the alignment as early as possible," said Ralph. "The drum is still right of the nose, and you haven't done anything about it." I eased the cyclic stick slightly to the right, and applied a small push to the right rudder pedal.

"Target range five…four…three…two," called Bob. By then, I had brought the Whirlwind into a hover. I was concentrating on the sea-hover techniques which Ralph had taught me.

"Target still range two, now slightly left of the nose," said Bob.

"That means you haven't quite made it," said Ralph. "You entered the hover a wee bit too early. We never got to the target's over-head."

"Target now range three, drifting further left," said Bob.

I applied a firm input to the cyclic stick.

"Relax. Relax!" said Ralph. "The drum has a mind of its own. Don't start over-controlling."

"Target still left of the nose, range two," called Bob. I looked at the horizon; it was too indeterminate to offer much guidance. My eyes then searched for clues closer to the helicopter. In the vicinity, I noticed a patch of sea-weed. I tried to use it as a hover reference.

"That sea-weed is a help," said Ralph. "But look for other possible references as well."

"Target range one…*steady…steady,*" called Bob. The 'steady' call meant that we were directly over-head the drum.

"Steady…steady. Down two." We were slightly too high, and Bob had asked me to descend.

"Your height is good. Left one," said Bob.

I continued to follow Bob's instructions. The call I eagerly awaited was: "Up gently." It meant that the winch hook had grabbed the rope surrounding the drum. At that point, with a slight upward movement of the collective lever, we would have lifted the drum clear of the water. Simultaneously, Bob would have winched-in the drum to keep it clear of the water's surface. Regrettably, however, the call never came. Bob persevered with his commands. There were endless small corrections. It became quite wearisome and prolonged.

Eventually, Ralph said: "It's an awkward kind of a day. I'll give you a break." As I handed control of the Whirlwind to Ralph, the helicopter suddenly seemed to come to heel. The master was in control; his authority had to be obeyed.

Quite quickly, we heard the call: "Up gently."

Ralph looked across at me. He must have anticipated my look of despondency.

"Don't worry about it," said Ralph. "Next week we'll be looking at 'mountain flying'. That's when the problems really start."

CHAPTER 30

Into the Mountains

"I've done a few rescues in these mountains over the years," said Ralph. "It can get pretty hairy at times."

We had just flown through the Llanberis pass. We had passed the castle to the north of the town of Llanberis; it appeared sombre in its position of over-view. Mount Snowdon towered above us on the right side of the Whirlwind helicopter.

"We'll follow this road for a few more miles," said Ralph. "Then just before the next road junction, there's a small lake in the hills. It'll be a good spot to start our training."

As the road junction came into view, Ralph suggested that we reduce our airspeed. "Bring your speed back by about 15 knots," he said. "It'll be easier then to find the gully leading us to the lake."

Below the Whirlwind, cars were driving on the road; a few flashed their lights when they saw the aircraft. The visibility was good; the features in the stark landscape showed up conspicuously. We felt some turbulence in the gusty conditions. "The wind does strange things around these mountains," said Ralph. "You have to be on your guard all the time."

We had just passed the road junction, and I turned the helicopter left on to a northerly heading. "The gully we're looking for should appear quite soon," said Ralph. I had noticed a stream coming down the mountain side. The area was covered in deep scars where running water had carved various gullies and ravines.

"That's the one I'm after," said Ralph. I turned the Whirlwind further left as we followed the line of the gully. We left the road behind us. The aircraft gained height as we delved into the mountainous terrain. Turbulence became more pronounced and unpredictable.

The craggy rocks and slate of the land below us mingled with assorted peat and moss.

"There's the lake just coming into view," said Ralph. The small lake was surrounded by high ground to the south and west. Further south, the peak of Mount Snowdon was occasionally sighted.

"Maintain about fifty knots airspeed," said Ralph, "and fly around the perimeter of the lake. We'll just get a feel for the wind, and the general conditions."

As I flew around the lake, a few birds were disturbed. "At this speed," said Ralph, "the birds generally have enough time to avoid us. Nevertheless, be aware of the potential for a bird-strike." Following the edge of the lake, I turned the helicopter away from the high surrounding terrain. "This lake isn't one of them," continued Ralph, "but you have to be careful; some of the lakes around here are bird sanctuaries."

We approached our start-point on the scenic tour around the lake. "What do you reckon is the wind direction?" asked Ralph. I thought for a moment.

"Well," I said, "it's all over the place actually."

"Exactly," said Ralph. "That's our problem in a nut-shell."

We continued to fly along the edge of the lake. "Look at that high ground above us," said Ralph. "The general wind direction today is westerly, blowing from the sea. But as the wind comes down the side of the hill, down-draughts and complex eddies are formed. It's a big problem for helicopters."

❖ ❖ ❖

Falklands, March 1986. Eight years in the future. Flying a Sikorsky S61 helicopter, I had departed the Falklands Island settlement at Fox Bay. The helicopter was buffeted by the strong gales. Our passengers looked pale and anxious. With the wind behind us, we quickly reached a site on the southern tip of West Falklands. We dropped off some troops and their equipment, and then took-off for our next destination: the settlement at Port Stephens. On that leg, the aircraft was soon approaching high ground. We were flying directly into the wind on a generally northerly heading.

"Our ground speed is very low," said my co-pilot. We were flying at 2000 feet, just about level with the top of the approaching high ground.

"We're already at maximum airspeed of 110 knots," I said. We were slowly creeping towards the high ground; the settlement was just beyond. The Falklands War had ended four years ago, but there was still a strong presence of British troops. I was a civilian pilot by then, and my Company had the contract to fly the military and their equipment around the Islands; Port Stephens was a regular place on our schedule.

"I'm not so sure we're moving," said the co-pilot eventually. "Look below us." He was right. Looking directly below the helicopter, it was clear that not only had we stopped forward flight, but we were actually starting to move backwards.

"This is incredible," I said. "Here we are at maximum power and maximum speed, quietly going backwards." I looked across the cockpit at the co-pilot. He shrugged.

"Well, in the absence of any other ideas," I said. "I suggest we turn down-wind to get away from this airspace. We'll then try the approach at 3000 feet. Hopefully that'll put us above the effects of this down-draught."

As I turned the S61 downwind, it was alarming to see how quickly the Falklands gale took us away from our intended destination. We ended up several miles downwind. Eventually, I turned back on to a northerly heading, facing into the wind again.

In the clearer airspace, we had the power to climb to 3000 feet. Our action had been unconventional, but effective.

"That's more like it," said the co-pilot. "At least we're making progress now." The helicopter was no longer under the influence of the overwhelming down-draught. With relative ease, we flew over the peaks of the hills. Once past the hill-top, we were beyond the effect of the down-draught. We had no problem approaching the settlement at Port Stephens.

"That was freakish," I said to the co-pilot after landing. "I haven't come across down-draughting like that before."

"Nor me," he said.

❖ ❖ ❖

February 1978. "We'd probably have difficulty entering a hover here," said Ralph. "It could be dangerous even to attempt." We looked at the high ground towering above. "So we'll now move away from this area. Let's see how it feels on the other side of the valley."

As we flew away from the lake, I turned the Whirlwind to follow the line of the contour towards the opposite side. Quite quickly, I noticed that we needed less power. The Whirlwind felt more comfortable, in spite of the ever-present turbulence.

"That's proved the point," said Ralph. "So we'll continue in this direction to the next valley. We'll come across a whole new set of problems there."

I tried to stay in the 'up-draughting' air as we proceeded to the neighbouring valley. We were in a remote outback. Wildlife emerged from time to time, alarmed by the helicopter's noisy presence. As we entered the new valley, a sheer rock-face soon became evident.

"What do you make of that, then?" asked Ralph.

I examined the scene. "The rock-face looks to be in up-draughting air," I said.

"It looks likely," said Ralph. "As a first step, we'll make a 'dummy' approach to that spot near the base of the rock-face. If you don't like it, then overshoot towards the clear area away from the high ground."

The Whirlwind continued to be buffeted by the turbulence in the area, but creeping along the base of the hillside, we were not inhibited by lack of power. We seemed to be in 'up-draughting' air as I made the 'dummy' approach, and came to a hover.

"There's no problem with the power here," I said to Ralph. Above us towered the dramatic rock-face.

"OK," said Ralph. "Imagine there's a survivor half-way up. Could we get to them?"

My eyes looked upwards, following the line of the rock-face. Towards the top of the hill, the rock jutted out. It looked awkward and dangerous to attempt.

"From the point of view of the power available, we could probably attempt it. The danger of a 'tip-strike' would limit us," I said. The striking of the blade tips against the rock's surface would be potentially disastrous.

"I agree on both counts," replied Ralph. "We haven't got a crewman or winch operator on board today, so we can't attempt getting close to the rock-face. Even with the crewman to guide us there would be problems."

❖ ❖ ❖

August 1997. A Sunday, nineteen years in the future.

"No further right, Skipper. Standby." The winch operator hung out of the Sikorsky S61 Coastguard helicopter as far as possible. He was secured by his harness, but nevertheless felt vulnerable. The winch operator was straining to see the area around the helicopter's tail rotor. The Coastguard had received an emergency call for help. A climber had fallen on the Stac Pollaidh mountain, 2009 feet high, in Wester Ross, Scotland.

"How are we doing for power?" the winch operator asked the aircraft Captain.

"Not too bad; I'm happy to continue," replied the Captain.

The winch operator looked around the area to the rear of the helicopter once more. Then he called: "Move right two, Skipper." The Captain eased the helicopter in the requested direction. The helicopter blades looked perilously close to the adjacent rocks. The winch operator endeavoured to keep one eye on the casualty below, but his main concentration was on the ever closer-looking rocks.

"We'll still not quite reach the casualty from here," called the winch operator.

"Roger," said the Captain.

"Move back one, Skipper," said the winch operator. "Easy does it."

The Captain strived to maintain a constant relative position to the rock-face as the helicopter was buffeted by turbulence. Slowly and skilfully he moved the helicopter marginally backwards. The tail rotor was even closer to the rocks.

"Steady. Steady." Called the winch operator. Firmly gripping the handle above his head, he again swung himself out of the cargo-door opening of the S61. He faced a deepening dilemma; the machine was too close to the rocks for comfort. However, his Captain was highly experienced, and the casualty desperately needed their assistance. At what point should he call a halt?

"I'm going to take you a bit further right, Skipper," said the winch operator eventually. "Standby."

The Captain continued to hold the helicopter's hazardous position. Once again, the winch operator craned his neck to the rear of the machine.

"Move right one, Skipper," called the winch operator.

Slowly, almost painfully, the Captain followed the winch operator's instructions. So far so good.

The winch operator peered below as he strained to further analyse their situation. The Captain held the S61 in the hover. Suddenly an extra strong gust of wind struck the helicopter. There was a loud crack. The S61 started to vibrate violently.

"Tip strike! Tip strike," yelled the winch operator.

"We're going down. Brace…brace…brace," shouted the Captain. He struggled to maintain control of the helicopter as it shook furiously. Unusual, unnatural forces were affecting the flying controls. The S61 was descending rapidly. The winch operator hurled himself inside the cabin, and attempted to strap himself to a seat.

The Captain had limited control available to him by then; the rotor tips had been severely damaged. He persisted to struggle with the machine as it descended. The Captain pointed the helicopter at a clear area by the base of the hillside. The S61 continued to judder savagely during its downward journey. The rate of descent was far greater than normal.

At the last moment, the Captain raised the collective lever to cushion the landing. There was a crunch as the helicopter crash-landed heavily.

The Captain rapidly closed down the helicopter's engines. He looked around him grimly. The whine of the engines diminished. There seemed to be an eerie calm for one moment. The shock of their predicament was still being absorbed.

Hastily, professionally, the winch operator made for the fire extinguisher. He unclipped it, and held the bottle in his hand. "Better evacuate quickly in case of fire," called the winch operator. His words galvanized the crew into action. They scrambled out of the damaged helicopter, and moved to one side.

It could have been worse: the soft peaty soil had cushioned their impact. As the crew surveyed the dismal scene, they had one overriding thought: at least all those on board had survived the crash.

❖ ❖ ❖

February 1978. "We'll leave this rocky area for now," said Ralph. He pointed to the left side of the aircraft. "Head for that col over there. We'll over-fly it on our way back to base."

As I flew towards the col, the turbulence caused by the steep rockface diminished, although conditions were still blustery. The peak of Mount Snowdon appeared again from time to time. I noticed the effect of the wind on our ground speed as we approached the dip in the hill-tops.

"There are going to be some strange effects as the wind is channelled through," I said to Ralph. I re-checked the aircraft's ground speed, and slightly raised the collective lever. Then I adjusted our heading.

"That's the idea," said Ralph. He glanced at me, and grinned. "We'll make a Search and Rescue pilot out of you yet," he gibed.

CHAPTER 31

Learning a New Language

As I stepped out of the car, I stretched my arms; the drive from Valley had taken several hours. Just before leaving there, I had been given some news. When I entered the house, I called out: "It's meee!" The family atmosphere was all about; it was such a fine feeling. Two blond-haired children bounced up to me. Their mother followed closely; there were hugs as the family re-united.

"I've got some news," I said eventually. Sue looked at me. "The Posting Notices are just in: we got our first choice."

"You mean we're going to Germany?" asked Sue.

"We sure are," I replied.

"That's grrreat," she said.

"What's happening?" asked Lizzie.

"Why," agreed her brother.

The family Atlas was produced, and we turned the pages impatiently. "It's called 'Gütersloh', and is near a place named 'Bielefeld' in northern Germany."

At length, I said: "We must go and tell Tom and Maggie next door. They were there with me last time, ten years ago."

Maggie was an attractive north country lass with tangles of red hair. She and Sue had quickly become firm friends. Maggie's husband, Tom, was an Air Traffic Control Officer.

"Fee, that's great," said Maggie when we told her our news. "You'll have a ball! Lucky things."

"Tell me all about Germany," Sue said to Maggie.

"Eee, it's cool, it is!" said Maggie. "How's your German?"

"Not too good," replied Sue.

"Richard will teach you," said Maggie. "He can speak the lingo."

"The thing to remember," said Maggie, "is to go around the place bold as brass. They can be quite snooty over there. Don't you worry about it. Remember who won the war. Don't be put off, now."

"As in 'farce', don't you mean 'brass', Maggie?" I enquired.

"Eeee, don't be a prat," said Maggie. "As in 'lass'…I mean 'brass'."

I glanced at Maggie, who was grinning slightly.

"You southerners. You haven't got a clue," Maggie pointed out. Then she said: "Anyway, you'll enjoy the 'bratties' over there."

"Bratties?"

"Aye, bratties. *Bratwurst mit frites*. Sausage and chips. Delicious!"

"And what's the weather like there?" asked Sue.

"You get foggy days. The muck from the *Ruhr* tends to blow across Gütersloh and that area. It can be quite unhealthy sometimes."

"And what about travel?" enquired Sue.

"Aye, its great. You can go all over the place. You can drive to Hannover. There's skiing at Winterberg and Villingen. You could go to Berlin on the military train. Then there's good access to Denmark, and to the Low Countries, and there's always southern Germany."

Maggie thought for a moment. "Then the Mess over there is really cool," she said. "What's 'is name, boring Goering, or someone, in the War built this 'falling beam' thingy. He would tank-up his Luftwaffe Officers with beer, and then release a secret catch. A ceiling beam would then fall down half-way, scaring the daylights out of them all. It still works. It's great."

"We're not going for a few months yet," said Sue. "Richard's got to do the Wessex course at Odiham first."

"Eee, don't you worry about it luv," replied Maggie. "I'll look after you when he's away." The family would remain in our home at Shawbury for the duration of the Odiham course.

In June 1978, I duly joined Number 240 Operational Conversion Unit at Odiham, Hampshire, for a three month course.

My Instructor, Bob, was a tall individual with many flying hours' experience on the Wessex helicopter.

"We'll start by looking at 'sloping ground' work," Bob said to me.

For 'sloping ground' work, the Wessex had to be flown towards a suitable gradient which had been selected for training flights. The approach to the side of the hill was made cautiously, and a wind assessment had to be made bearing in mind any local effects. If it was too steep to land, then one wheel would be placed tentatively on the ground. Then the other main-wheel, and the tail-wheel, would be kept airborne. The pilot was hovering, in effect, just part of the helicopter.

"That's the idea," said Bob. "Hold it here for a few minutes to get the 'feel'. It can get problematic, especially in gusty conditions. But you may need to hold it like this for a while as troops unload and load."

The atmosphere at Odiham was different from other units I had known in the Royal Air Force. This was the centre of the 'SH', or 'Support Helicopter', world. It was more like being in the Army than the Air Force. Green tents and camouflaged vehicles littered the airfield. There were soldiers carrying bulky radios; the men used slow and deliberate language as they repeated a constant flow of messages. Wessex helicopters were scattered around, seemingly at random. The aircraft types I had been used to flying were always at the centre of attention. Here, the helicopters appeared to be treated with as much priority as the prolific number of four-ton trucks. There was a basic, earthy, stench everywhere. "This is the back-end of the Air Force here," an Officer remarked one time.

"You'll get lots of practise at the 'sloping ground' work," Bob had said eventually. "That'll do for now. We'll look at some 'quickstops' next," he said. "Let's fly over to Upavon for this."

We crossed the M3 motorway, with the town of Hook just north of our flight path. The motorway was busy with an endless stream of traffic. On our left was the sprawl of Basingstoke; even from the air the sense of suburban over-crowding could be felt. It was a different scene to the openness we had enjoyed in Scotland and Wales.

The Wessex headed due west for some 25 miles towards the small airfield at Upavon. "It's much quieter here," said Bob as we approached the airfield. "And we'll have a bit more room." He backed the Wessex into one corner of the airfield in preparation for our 'quickstop'.

"OK. I'll demonstrate one, then you can have a go," said Bob.

He lowered the nose of the Wessex by placing the cyclic stick forward. Then he pulled up the collective lever. With the application of collective pitch, he stopped the machine from descending. The helicopter began to move forwards, and the airspeed increased rapidly.

"Right," said Bob. "We've got 90 knots indicated, so I'll commence the 'quickstop'." He simultaneously lowered the collective lever, and pulled back on the cyclic stick. The aircraft nose reared into the air. The airspeed reduced rapidly. "I'm making constant small adjustments with the controls," said Bob, "to prevent a climb or descent. Our main aim is to stop as soon as we can, in the shortest possible distance." As the helicopter's airspeed fell to zero, Bob brought the machine into a hover.

"Any questions?" Bob asked me as he held the Wessex in a steady hover.

"No thanks," I replied.

He then handed the aircraft controls over to me. "You have to be really careful about 'vortex ring'," said Bob, "especially when you do downwind 'quickstops'. But we'll look at those later. For now, you can hover-taxi back to where we just started." I pushed the rudder pedal to turn the Wessex through 180 degrees, and slowly hover-taxied to the start point for our 'quickstop.'

"Looks all clear," said Bob. "Off you go."

As I pushed the cyclic stick forward, the nose of the helicopter tended to drop promptly. "Anticipate that with the collective lever," said Bob. "Operate the two controls in a more balanced way." The airspeed built-up briskly: 20 knots…30 knots…40 knots. Soon we were at the required 90 knots.

"Quickstop…quickstop…*go!*" called Bob.

I hauled back the cyclic stick. I lowered the collective lever as far as it would go. The nose of the Wessex rose high in the air. "Don't climb," said Bob. "Use the cyclic in this situation to control your height." I made a cyclic correction. As the airspeed diminished, I brought the helicopter to a hover.

After a few more practices, Bob said: "Not bad. They're coming on. We'll have a look at a downwind 'quickstop' now. Remember what we said about 'vortex ring'."

'Vortex ring' was a potential problem peculiar to helicopters. In conditions of low speed, high rate of descent, and with power applied, loss of control was likely. A number of helicopter accidents had been caused by the phenomenon. The latter stages of the flare during a downwind 'quickstop' was a classic occasion when 'vortex ring' could occur.

"Just turn on the spot here," said Bob. "Then when it's clear, take-off towards the far corner. Remember the wind's behind us this time." As I did the spot turn, it felt awkward holding the machine in a downwind hover.

During the take-off, I was aware that the groundspeed picked up quickly, but the airspeed was slow to build up. The effect of the wind in our rear quarter was dramatic. The far corner of the airfield appeared to loom up quickly.

Then Bob called: "Quickstop…quickstop…*go!*"

I applied a high angle of bank to turn the helicopter into the wind. I lowered the collective lever, and simultaneously pulled back the cyclic stick. With the wind still behind us, the groundspeed was faster than before. As the airspeed was coming back, I raised the collective lever. Suddenly, the Wessex began to judder violently.

Immediately Bob called: *"I have control…I have control."* He rapidly took over the helicopter's flying controls, and pushed the collective lever down again. He continued the turn into wind. The juddering through the airframe began to decline. When we were pointing into the wind, he progressively reduced the helicopter's airspeed.

"That was a good example of incipient 'vortex ring'," said Bob. "You've really got to watch it. Another situation when you're likely to have problems is in a vertical descent, for example during instrument flying. You must ensure you have positive airspeed before raising the collective lever."

We practised some further downwind 'quickstops', then Bob declared that it was time to return to Odiham.

"Some good lessons learnt there," said Bob during the sortie de-brief. "It's all rather different from your fixed-wing life, I guess."

"You could say that," I replied.

"Anyway, its time to go home now," said Bob. "The week-end is upon us. Next week, we'll be looking at operating in confined areas. We'll also do some night flying, and demonstrate under-slung load work," said Bob.

After Bob's de-brief, I went to my car. The drive from Odiham to Shawbury took about three hours. Approaching Shrewsbury, the high ground in the distance signified the freedom of the Welsh hills. It seemed a heartening sight after the over-crowding of the south-east of England.

When I eventually reached home, the children were playing outside. They raced up to greet me. Then, as I walked from the car to the back-door, I became aware of German speech: *"Er ist nicht sehr teuer, aber er ist zu teuer fur mich."*

My wife's back was turned as I went in. She and Maggie were concentrating on a book. Suddenly, Sue turned round, and said: *"meine tasse ist zu klein!"*

"Hi gorgeous!" I said. Sue gave me a hug. "Hi Maggie!" I said.

"Impressed?" Sue asked as she pointed to a tape recorder. The machine continued to pronounce flawless German sentences.

"Sehr gute," I said.

"Have a good week?" the ladies had reverted to English.

"Aye," I said. Maggie looked at me sideways. "Better leave you two to catch up," she said.

"No, no. Don't rush off, Maggie."

"Its OK. Tom's due back now anyway."

"Well thanks for keeping an eye on the troops, Maggie," I said.

"No prob," said Maggie. "It's a pleasure. Anytime."

"By the way," said Maggie as she was leaving. "Your Alan. He's a right little monkey is that one."

"Why, what's he done this time?"

"Eee, he's great really. But look at him now, tearing about on that grass."

I grinned at Maggie: "Tearing about where, Maggie?". Maggie gazed at me disapprovingly. Her brow furrowed as she scowled slightly. Then she seemed to have a sudden change of heart. Her eyes twinkled as she looked in my direction.

"As in 'ass'…on the 'grass'!" said Maggie.

A Westland Wessex helicopter.

CHAPTER 32

To Germany Again

"What do you make of our Sandy, then?" asked George. We had just taken-off from Gütersloh, Germany, in a Wessex helicopter. George was the local Qualified Flying Instructor; he was flying with me on an area familiarisation flight. It was September 1978. I had finished the Odiham course, and had been posted to Number 18 Squadron. My family would join me when the 'Families' Officer' at Gütersloh had allocated us a Married Quarter.

"Not overly inclined to take a lot of crap," I said in reply to George's question. We were discussing the Commanding Officer of my new Squadron.

"You're not wrong," laughed George. "He hasn't been here very long. Already he's known as 'The White Tornado'!"

We flew at a height of 50 feet towards a helicopter low flying training area just north of Gütersloh, known as 'Paladin'. George opened the cockpit window on his side of the Wessex, and lit a cigarette.

"Like one?" he asked.

"No thanks," I replied. George put his foot up by the base of the open window.

"Yes," George sighed. "A weird one is that." A heavy smoker, George tragically died of lung cancer just a few years after the end of his tour at Gütersloh.

The White Tornado had been a subject of discussion at Odiham too. "He's a human dynamo," said one Officer. "The Air Force flows through his veins; I'm surprised there's any room for the blood. Mark my words: his aim is to single-handedly raise the 'Support Helicopter' world from the rump to the forefront."

In spite of the dire augury, in fact I had found the White Tornado to be friendly. With slightly balding fair hair, he would march about the place at a double-quick pace. Behind the fierce no-nonsense approach lay an astute mind. Also in the back-ground was a charming, intelligent wife. The White Tornado expected miracles from his fellow Officers. In return, he would back them to the hilt if anyone fell by the way-side.

As we approached 'Paladin', George said: "This place looks innocuous enough, and it's not even very big. But all the features are so alike that's it's amazingly easy to get lost." George had pre-planned a navigation exercise for me.

In the cabin of the Wessex was Frank. He acted as our crewman. I would get to know Frank well over the next two years. He was an Irishman, slight in build, with a practical and capable mind. When I knew him better, I thought of him as the 'Pied Piper': his kindness to children was striking, and quickly sensed by the younger generation. Normally the crewman would help with navigation, but on this occasion George had briefed Frank to leave it to me.

"It's important to stick to a height of 50 feet," said George. "I know it's lower than you're used to, but that's the *modus operandi* over here. Use the radio altimeter to check your height from time to time. It'll help you get used to the feel of an accurate fifty feet."

I had begun the first leg of the navigation exercise. The small town of Harsewinkel appeared on the nose of the aircraft.

"Considering the average height of the trees around here is about fifty feet," said George, "it tends to make the navigation quite interesting."

Beyond Harsewinkel was a large wooded area. George's route took us in that direction.

"Keep your airspeed at 90 knots," said George. "You're tending to allow the speed to reduce. I know it makes the navigation a bit easier, but for operational expediency we fly at 90 knots."

We were approaching the forested expanse; at a height of fifty feet, we were close to the tops of the trees. At that height, the wooded area appeared to extend to the horizon. An occasional track cut through the trees, otherwise there was a stretch of uniformity. George had planned the navigation exercise to test me. Just a short distance off track, and the turning point would be easily missed.

"I reckon we should see our turning point ahead at any moment," I said hopefully. George said nothing, but I noticed that he appeared to be grinning to himself.

"Well that's strange," I said after a bit. "I'm sure we should have come across the feature by now." George was nodding.

"Perhaps we've passed it," I continued. George was still nodding.

"I'll turn onto the next heading anyway," I said.

"Good plan," commented George.

We still flew over the large expanse of woods. The horizon appeared indistinct; 'The muck from the Ruhr' as Maggie had so aptly described it. I began to feel more uncomfortable. The navigation exercises we had practised at Odiham had been relatively straightforward. I had become reasonably adept at managing the 50,000 scale maps. I had left Odiham with the impression of being fairly competent at low level navigation. Quite quickly, I was becoming disillusioned. The thought that we were still just a stone's throw from Gütersloh made it seem worse.

Eventually, I felt that enough was enough.

"Sorry George," I said. "But I think I'm lost!"

George started to laugh. He seemed quite laid-back about it. He lit another cigarette.

"Don't look at me," said George, dragging on his cigarette. "I'm bloody lost too!"

I studied the map ever more closely. A worried frown marked my forehead.

"That thing's not going to help you any more," said George as he pointed at the 50,000 scale chart. "The best thing we can do now is cheat."

I looked at him in surprise.

"Just back there," said George, "was a road sign. You're in a helicopter now, not an 'aluminium death-tube'."

I brought the Wessex to a high hover, and turned back towards the road sign. Frank read out the sign: *'Clarholz...4Km'*. He had been following our progress on his own map.

"I suggest we head 035 degrees from here," said Frank. "It'll cut the corner to the next planned turning point." Frank winked as I nodded to him gratefully.

We flew over the wooded area once more as I followed Frank's suggested heading. I was searching for a small pylon which was marked on our 50,000 map. Again, I had an uncomfortable feeling as we approached the anticipated spot. I glanced at Frank; already I felt that he had become my ally. But Frank had a bothered mien as he studied his own map. He looked at me and shrugged. George appeared impassive.

"Well, going by our time," I said eventually, "I reckon we should have passed the pylon by now."

"I didn't see any pylon," said George.

"Nor me," Frank chipped in.

"Well either we're well off-track, or there isn't any pylon," I suggested.

"You can't always trust the accuracy of the maps over here," George pointed out.

"In that case," I said, "I'll turn now anyway."

"Fair do's," said George.

We were flying at the edge of the wooded area at that stage. The outline of the trees at least made a more distinctive feature. I felt a little more confident. George's navigation route was taking us due west at that stage, heading towards Münster, the capital city of the district of Westfalia. Shortly we would leave 'Paladin'. Ahead lay some large electricity pylons.

"What are you going to do about those pylons?" George asked me.

"Climb over them, I guess," I replied.

"They're around 100 feet high," said George. "So that wouldn't be very tactical."

George noted my look of confusion.

"We could go *under* them," said George. Such thinking was heresy in the United Kingdom.

"You're in Germany now," remarked George. "We do things differently over here."

As we approached the pylons, George said: "I have control." He brought the helicopter to a hover as we descended to a height of ten feet.

"Ready Frank?" George asked.

"OK Skipper," replied Frank. "You're clear below for light contact." George lowered the Wessex in the hover until the wheels were just touching the surface.

"Clear out Frank," called George. Frank unplugged his head-set, and stepped out of the helicopter's cabin. He then went ahead of the Wessex while George held the machine in a hover. Frank walked under the wires between the electricity pylons. He inspected them, and assessed their height. When he had walked to the other side of the wires, Frank held up both his hands above his head.

"Right," George said to me. "That means Frank is satisfied with the height clearance, and general safety aspects. What he'll do now is marshal us under the wires. I'd like you to do the flying. Obviously, there's very little room for error in something like this, so watch it! Pay particular attention to any height adjustments Frank may indicate. If you're unhappy at any stage, just land on the spot."

George handed me the flying controls, and I held the Wessex in the hover. George then indicated a 'thumbs up' signal out of his cockpit window. Frank had a final look around, and then began waving his arms in a steady action.

Cautiously, I eased the cyclic stick forward to start a slow hover-taxi. Frank continued to wave his arms as we approached the first set of electricity wires. The pylons either side towered above. The helicopter blades appeared to be just inches from the approaching wire.

"Don't look up," said George. "You'll only frighten yourself. Concentrate on Frank's marshalling." Frank continued to wave his arms. The wheels of the Wessex were very close to the ground. Our margin for error appeared to be alarmingly slim.

Soon we were directly under the first set of wires. I looked intensely at Frank as he signalled rhythmically.

Suddenly, Frank held both of his arms out in a horizontal position. Immediately, I stopped our forward movement, and maintained a hover. Frank then slowly indicated a downward motion with both of his arms. I eased the Wessex down into an even lower hover.

"That's the idea," said George. "This one's quite tight."

Frank then held his arms level again. He stepped backwards slightly, and then resumed his arm waving in a regular motion. I began to creep forward once more. The first wires were going behind us; soon we were approaching the second set of cables.

"Just because the first lot of wires gave us no problems," said George, "doesn't automatically mean we're clear under the second set. The lowest point under each set of cables is a variable feast. The other thing to bear in mind is the experience level of your crewman. Frank's been doing this a long time; he knows what he's up to. Some of the younger crewmen have to be treated a little more warily."

I continued to creep the helicopter forward under Frank's guidance. As we approached the nadir of the second set of wires, Frank suddenly held out his arms again. I brought the helicopter to a hover in an instant. I glanced across at George momentarily. He still held his normal impassive expression, nevertheless I sensed that he had become slightly tense.

Frank indicated a further downward movement with his arms. As I took the Wessex into an even lower hover, I was aware that one of the machine's main wheels had fleetingly touched the surface below.

George gave a brief shrug. "If necessary we'll ground-taxi this thing under the wires," he said.

Frank had started to wave his arms again, but I noted that the motion was measured. Slowly, I began to move forward. I did not feel the helicopter's main wheels touch the surface again, but it must have been a close thing. The low-point of the second set of cables was just approaching above our heads. The futile idea of ducking momentarily crossed my mind. Frank maintained his steady marshalling signal. The wires slowly began to move behind our heads.

"You can't relax just yet," George reminded me. "There's the tail rotor still to go."

In spite of the composure in George's manner, nevertheless, as I crept the Wessex forward for the final stages, I sensed a feeling of relief from the aircraft Captain.

Soon we noted that Frank was signalling less ponderously, and he took some backward steps as he pointed to a landing area.

"Well done Frank," said George as our crewman climbed aboard, and plugged-in his head-set.

"That second set of wires looked quite interesting," said the crewman. Frank was adjusting his harness in the cabin of the Wessex.

"I thought you looked a bit anxious at one point, Frank," I said.

"Me? Never!" he replied.

I gazed at Frank for a moment. He grinned as he made a 'thumbs-up' signal. Then Frank revealed his other hand. I soon saw that he had two fingers firmly crossed.

CHAPTER 33

The White Tornado

"Well you'd better come with me, then." The White Tornado had just spoken. A field exercise was planned for the following week, and the Squadron Planning Officer had been uncertain what to do with me, the new-boy on the unit. "You can sit in the left seat of my aircraft, and be my co-pilot," said the Commanding Officer.

"And we'll have Frank as our crewman," added the White Tornado. "You seemed to get on well with him yesterday. He can teach you how we operate in the field." After his pronouncement, the White Tornado went charging off.

When the start of the following week duly arrived, we had early briefings. The weather man predicted showers and gusty winds, with a band of more general rain later in the day. Then the Planning Officer outlined arrangements. Unlike some of the exercises at Gütersloh, which were initiated by a siren (usually sounding in the middle of the night), this exercise had been programmed in advance. The exercise, code-named 'Gryphons Galore', was planned to last for ten days.

As the aircrew left the Briefing Room, there was hectic activity all around. Wessex helicopters were lined up at the front of the dispersal area. Engineering staff inspected and serviced the machines. Four-ton trucks and various Land Rovers were being loaded with equipment. Soldiers and airmen moved purposefully to prepare for the days ahead.

The Squadron was broken into various 'Flights' for operating in the field. Each 'Flight' was autonomous, with allocated aircrew and engineers, a mobile operations unit, and independent engineering facilities. There were also separate domestic set-ups which included catering facilities, security arrangements, and the inevitable camping paraphernalia. It was a highly effective organisation. The whole regime was geared to provide flexibility, and a powerful operational capability.

The White Tornado was in his element. He galloped around at an even faster pace than normal, issuing instructions at every opportunity. He did not like to see inactivity, and had a knack of spotting signs of slow progress.

"Why wasn't that done over the week-end?" he asked a junior engineering Officer.

"Hope you'll do better than on the last exercise," he suggested to an airman field-caterer. "If half my Squadron go down with the squits again, I'll hit the roof!"

"That'll make a nice change," the cook said to his mate as the White Tornado raced off.

When he came across a crewman who had pleased him, the White Tornado said: "Good man. Well done! Get ahead of the game, that's the idea."

Eventually, when the time for departure into the field had been reached, the White Tornado said to Frank, and to me: "If you two get the machine turning and burning, I'll join you shortly."

Frank and I walked out to our allocated Wessex: XR 505. We carried bulky rucksacks, and sleeping bags which dangled below the rucksacks. The gear would provide the basis of 'home' for the next ten days. Frank secured the equipment in the back of the Wessex, while I walked around the machine performing external checks. Then I climbed into the cockpit, and went through the drills to start both engines and the rotors. We were soon 'burning and turning'.

The White Tornado seemed to be in a good mood as he joined us. "OK Richard my boy," he said. "You do the flying, and Frank will navigate us there." Most of the other helicopters had already left, flying in formation to their various 'flight' destinations in the field. Our Wessex was part of the 'Headquarters Flight', and we flew on our own.

Soon, we had departed to the east of Gütersloh, and headed towards the 'Bielefeld Gap'. "That's the idea," said the White Tornado.

THE SPICE OF FLIGHT • 209

"Stick to fifty feet. It's a good tactical height, and apart from anything else will keep us clear of those pesky Harriers." Harrier vertical take-off aircraft from 2 Squadron and 4 Squadron were also deploying into the field as part of the overall exercise. "They're a pretty namby-pamby bunch," remarked the White Tornado. "We're the ones who really get on with the work around here. They spend half their time building over-complex 'hides', and the other half fussing about mud spoiling their paint-work." The White Tornado sniffed. "Whereas we are quick and mobile, and get on with the job. We can get an individual 'flight' packed-up and on the move within half an hour. Quicker if there's a 'crash-out'."

The 'Bielefeld Gap' was prominent as we approached the Teutoburger Hills. Passing through the 'Gap', tall radio and TV masts on the hill-top towered above us. To the north, the city of Bielefeld, with Herford beyond it, came into view. The Teutoburger Hills were covered in trees; dramatic swathes of early Autumn colour were just becoming evident.

"The next part of our route takes us through hilly areas," said Frank. "Then eventually we'll come to the flat lands around Hannover."

Crossing the River Weser, we saw the town of Hameln, famous for the Pied Piper legend, to the south of us. "You'll have to take your kids there when your family have joined you," said the White Tornado.

We followed a railway line which led into Hildesheim, at which point we spotted the expanse of flat plains encompassing the city of Hannover. We were heading for an area between Hannover and Braunschweig. I noticed banks of prominent electricity pylons stretched across the plains. "Don't worry," said Frank. "We won't be flying under all those wires on every occasion. It takes up too much time."

"You can get an over-view of the strategic position here," said the White Tornado. "The hordes of Soviet tanks would swarm across these plains in the event of World War 3. It's ideal tank country this." He made a sweeping motion with his arm. "Our job is to support the army by moving ammunition and troops around the theatre of operations." The city of Hannover was conspicuous as it stood in the surrounding countryside on our left side.

Eventually, Frank said: "Five kilometres to run, Captain."

"Thanks, Frank." By then, we had flown well beyond Hannover, and were becoming reasonably close to the border between West and East Germany.

We were making for a wooded area. In the centre of the woods was a clearing. We aimed to land the Wessex in the clearing, then we would join the set-up prepared by the advance party: the operations centre, the engineering facility, and the domestic site inside the woods. The natural environment, and the cover of the trees would provide ideal camouflage. The advance party of men and vehicles should have reached the pre-planned stop some hours before us.

"Three kilometres to go," said Frank. Ahead we could see the clear area within the woods. I made an assessment of the wind direction.

"Sounds about right," said the White Tornado. "Just carry on and land us in the clearing." I made a left turn into the westerly wind, and began to reduce the helicopter's airspeed.

Soon I said: "Your directions, Frank."

"Roger," replied Frank. "Your height is good, no lower. The clearing is on the nose range twenty-five. Your speed is good. Maintain this heading. Clearing now range fifteen, your height is still good." Frank was holding onto the winchman's handle by the cabin door as he looked around.

"Range now five…four…three…two…one. Steady. Steady," called Frank. He made a final inspection all about, then continued: "You are clear below. Down fifteen." I lowered the helicopter's collective lever to start our vertical descent. "Your tail is clear, as you come down. The trees to the left are fairly close, otherwise your position is good. Down ten. No further left. Down five…four…three…two…one…just touching." We felt the helicopter's wheels bouncing lightly on the forest surface.

"Hold it there," said Frank. He looked at the wheels of the Wessex. "It looks fairly firm. Clear to land." I fully lowered the helicopter's collective lever, and we felt the machine sink onto its oleo struts.

"Good. Splendid!" said the 'White Tornado. "I'll nip out here, and leave you to close down the machine." He climbed out of the cockpit, and charged off towards some vehicles hidden within the trees.

Meanwhile, Frank was standing in front of the Wessex to supervise the shut-down procedure. I closed down both engines, and waited for Frank to indicate 'droop-stops' in. This was soon shown by Frank: he pointed both thumbs together above his head. The 'droop stops' were designed to prevent the rotor blades from flapping around in high winds.

I applied the rotor brake, and as the rotors came to a halt, Frank returned to the helicopter's cabin. He produced some dark green webbing, and together we secured the webbing over the windshield. "We do this every time," said Frank. "The bulk of the fuselage is painted in camouflage colour, but a windshield glinting in the sun can give away our position."

Then we hauled the rucksacks onto our backs, and made our way towards the operations centre: a four-ton truck with a 'box' body. Beside the operations truck was another four-ton vehicle, which had been converted into sleeping accommodation for the White Tornado.

As Frank and I walked up the steps into the operations centre, we heard the White Tornados' voice as he spoke on the telephone.

"Yes, yes, yes!" the White Tornado was saying. "Talk, talk, talk. I need action, not more blather." The White Tornado fell silent for a moment as he listened again. Then he resumed his harangue: "Sod you, Bruce," he said. "You're as bad as the army. Couldn't organise a punch-up in a Belfast brothel." The White Tornado glowered as he slammed down the receiver. Suddenly, he beamed as he spotted Frank and me. He became urbane; he felt comfortable with his flying colleagues.

"Gentlemen," said the White Tornado. "Come in…come in. Have a cuppa tea. The makings are over there. I'll have one too, while I think of it. Looks like the camp site here's been pretty well prepared, so Frank can show you the whole set-up in a minute." The White Tornado nodded at me.

The White Tornado appeared relaxed as he pointed to his operations empire. "Monty would have been proud of this," he said.

Inside the 'box body' was an impressive arrangement of maps, books, instructions, and operations orders. Everything had been carefully considered, down to the last china-graph pencil. As it happened, when Field Marshal Viscount Montgomery of Alamein had retired, he lived in a large house near my uncle in Hampshire. My cousin Hew and I had been invited one time to the Field Marshal's home. We had looked around the caravans used by him as Commander of the Eighth Army during the World War 2 El Alamein campaign in North Africa. "I think Monty might have been quite envious," I said.

The White Tornado seemed contemplative as we sipped our tea. He offered me snippets of advice about living in the field environment. "Any idiot can live uncomfortably in this type of regime," he said. "It's the clever one's who survive, and keep themselves fit and clean. Look at our Marine Exchange Officer. His uniform is always smart, and he keeps his boots polished."

At that point, the telephone was answered by the operations assistant. "Sir," said the assistant, "it's for you." The White Tornado took the telephone. After a pause of a few nano-seconds, an explosive bawl shook the operations caravan as he bellowed down the telephone. Frank gestured towards the door, and the two of us slipped out.

"We'd best put up our tent for the night," said Frank, " before it gets much darker."

The two-man tent was in fact quite quick to put up and take down. We found a good spot reasonably near the Wessex, but under cover of the trees. Inside the tent, we assembled two wire framed beds. "These beds are actually fairly comfortable," said Frank.

As we walked around the rest of the camp area, I smelt the strong earthy scents which were to become so familiar over the next two years. The movement of men and vehicles had already churned up mud. The distinctive reek of damp camouflage netting was all around.

"That's the mobile water supply," Frank pointed to a large black tank set on a wheeled chassis. "The water inside is OK to drink. It's also used for washing. As the CO said, there's a lot of emphasis on personal hygiene. The medics make regular visits. Aircrew are required to get adequate rest. In the event of war, we'd be expected to keep going like this ad infinitum. It's quite a thought."

Leaning against the side of a four-ton truck, I spotted a row of spades. "Use those when you need the latrines," Frank pointed out. "Take a spade with you into the forest." He laughed at my look of disdain.

"Well, let's get some grub now," said Frank, as we wandered towards the Mess tent. "Then we'd better go to bed, I guess. It'll be an early start tomorrow."

In the Mess tent, we chatted with other members of the 'Headquarters Flight'. It was the time when people had a chance to relax. Some folk played cards; there was an endless supply of tea, and plenty of food. Eventually, after staying around the Mess area for some while, Frank and I walked to our tent. It had become quite dark. The forest echoed to the sound of falling rain. For tactical reasons, we used torches with green coloured lenses.

We laid the sleeping bags on top of the beds, and climbed inside.

"Sleep tight," said Frank.

"Thanks," I replied.

"And welcome to the weird and wonderful world of helicopters," sighed Frank.

My eyes stayed open as I reflected on the graphic change in lifestyle since my days as a fighter pilot. Just a year or so ago, I would not have believed this possible. The rain beat noisily on to the sides of our tent. It made a monotonous drumming sound. Soon, the nocturnal resonance's were joined by Frank's snoring.

I shivered in the cold night air.

CHAPTER 34

Border Patrol

I awoke with a start. It was still dark, and the rain continued to beat down. Frank was sitting bolt upright.

"*Stand to...stand to!*" An urgent voice could be heard in the distance.

"Better get up, quick," called Frank. He pulled on his flying boots. We were already part-dressed; we had slept in some of our clothing for warmth.

As I pulled back the flap of the tent, we were aware of the sound of feet running in the area of the clearing. The occasional flash from a green-lensed torch could be seen. In the background, the drone of a generator could be heard. Our eyes, still accustomed to the dark, made out fleeting images.

"Let's make for the helicopter," I said to Frank. "We can help defend it."

We crept through the trees towards the Wessex. Damp forest leaves brushed against our clothes. The ground felt soggy and saturated from the over-night rain. Another torch flash was seen; closer to us than before. We soon reached the edge of the clearing.

Suddenly, we heard the distinctive sound of 'blanks' fired from rifles. Then we heard raised voices in the distance; the noise of shouting drifted through the night air. I checked my watch: it was 4.30 am.

Frank and I made our way around the side of the clearing, towards the operations 'box body'. We moved slowly and cautiously, and kept checking for signs of intruders interfering with the Wessex. In spite of the rain, the gradually improving light signalled the progress of dawn. At that moment, we saw someone run up to the helicopter.

"That's Dennis," said Frank as he peered through the dim light.

"We'd better see what he's up to," I said.

We hurried from the edge of the clearing towards the cabin of the Wessex.

"Hello, you two," said Dennis. He seemed quite breathless. "The CO's just ordered a crash-out. We've been attacked by some SAS troops. It's just about light enough to get airborne now. You'd better get your kit quickly."

Frank and I scrambled back to our tent. We hastily packed our rucksacks, and dismantled the beds. Frank's experience of field operations proved invaluable as we nimbly took down the tent, and folded-up the soggy material. Within minutes, we were back at the helicopter.

Dennis was looking around the outside of the machine. "You'll be flying with me today," said Dennis. "The White Tornado is tied up with other things."

Dennis and I climbed into our respective cockpit seats. I sat in the co-pilot's seat on the left side. "If you don't mind," Dennis said to me. "As we're in a rush, I'll do the start-up and take-off." Dennis was a stocky, well turned-out individual. He was highly experienced on the aircraft type, and on operations in Germany. Frank stayed outside, monitoring our start procedure. He held a fire-bottle cradled in one arm – normal procedure in case of an engine fire during the start-up. The process was slick and efficient. Within moments, we were 'turning and burning'. Frank climbed into the cabin, and secured the crewman's harness around his waist.

"Clear above and behind," called Frank as he leant out of the helicopter's cabin, checking for obstructions on all sides. Dennis raised the collective lever, and began a vertical climb out of the clearing. Frank gave further instructions as we gained height. The trees scattered their covering of rain as the powerful helicopter blades beat the air. We caught glimpses of hurried movement as soldiers and airmen prepared for the 'crash-out'. The first vehicles were already on the move.

"We're heading for 'D' Flight initially," said Dennis. "We'll co-locate with them until the HQ Flight is established in its new location." He handed me a slip of paper with a grid reference. I read out the reference to Frank, and we marked the position of 'D' Flight on our 50,000 scale maps.

"It looks as if they're in a farm complex, about 15 kilometres from here," I said.

The visibility was quite poor in the damp conditions. We flew over an *autobahn* with an endless stream of headlights marking the early morning traffic. We monitored the FM radio in our aircraft in case of messages, but received none.

"Farm complex 4 kilometres just left of the nose," called Frank.

"Visual," said Dennis. He reduced the helicopter's airspeed as we approached. "We'll circle around the farm to work out where to land," Dennis said.

To one side of the farm buildings was a line of trees, which led into a small wood. By the line of trees, we soon spotted four Wessex helicopters dispersed to give maximum camouflage effect. An airman ran to the area behind one of the helicopters, with his arms held high.

"Marshaller sighted," confirmed Dennis. He obeyed the hand-signals of the marshaller, as Frank checked all around. Soon we had landed, and we spotted a figure walking towards us as Dennis closed down our machine.

"Gentlemen, good morning and welcome!" said Pete, as we climbed out of the Wessex. "Heard about your little *fracas* this morning. Come and have some breakfast, and then we've got some interesting tasking lined up for you." Pete was the Flight Commander of 'D' Flight. He was a friendly, good-humoured man. We followed him to the Mess tent, and became aware of the smell of fried bacon. "Now then," said Pete. "We pride ourselves on our good grub on this Flight. Feed the inner man, that's what I say. The chef's porridge is highly recommended. Then to follow, he does a mean eggs and bacon. Egon Ronay would be proud of him."

We carried our own plates and cutlery at all times, kept in the top of the rucksacks. In the mess tent, wooden tables and chairs had been laid out. To one side, the field kitchen was set up with mobile gas cookers and special stainless steel containers. At the end was an urn containing hot tea. We loaded our plates, and selected a table where we sat as a group.

"Well, what does the new pilot think of it so far?" Pete asked me as we ate our breakfast.

"Rubbish," interjected Dennis.

"Where did you guys sleep last night?" asked Pete.

"In our tents," I replied.

"Bad luck!" said Pete. "We chatted to the farmer, and he offered us a barn. There was fresh straw and everything. We had a really good night. The farmer has even allocated us a loo and wash-basin."

"All right for some," said Frank.

"It's the way forward," replied Pete. "These guys are only too well aware of their proximity to the border. Come WW3, their hope is that we'll remember their co-operation, and return. As a consequence, these are the farms that'll have some form of defence. From our point of view, we benefit by a higher standard of comfort, and the crews will be more fighting fit."

After breakfast, we followed Pete to the 'D' Flight operations vehicle. "Here are the weather briefs and the Flight Safety notices for today," he said, handing us sheets of paper. "While you study those, I'll check on your tasking." The weather for the day was forecast to improve markedly, following the passage of a frontal zone through our area.

Pete had returned with our tasking details. "Right," he said. "It looks like a mega-formation day today. We'll be operating near Braunschweig, quite close to the border. In the border area itself, we'll be guided by a Gazelle helicopter with Border Guard personnel on board. As you know, if we cross the border by so much as an inch, there's all hell to pay." Pete pointed to a map. "I'll show you in more detail on this," he said. "But first we'll round-up the rest of the aircrew, and I'll brief everyone together."

When it took place, the briefing was short and to the point. The aircrew marked individual flying maps, and noted their positions in the overall formation. Afterwards, the aircrew went to their respective aircraft to pack personal and other equipment on board before leaving. "Never get separated from your kit," was the adage we had learnt at Odiham.

Soon, all the helicopters began to start up. Ground crews were on hand in case of starting difficulties. The crewman from each Wessex stood in front of the machines, monitoring the start sequence. Then they returned to their helicopters, and strapped into their special harnesses. The crewmen stood on steps by the cabin doors, half-out of their machines, as they peered all around.

The lead helicopter then made a leisurely take-off, and the others followed in their pre-briefed order. Our Wessex, as the visitor, was the last in line. We followed the other four aircraft in a loose line astern formation. All the helicopters listened out on the FM radio, but nothing was said; the operations were generally in 'radio silence' for tactical reasons.

Soon, we approached a spacious field where other Wessex helicopters had arrived ahead of us. Our formation leader made a brief circuit of the field, and then landed next to the other aircraft. We followed in succession, and all the helicopters closed down.

After leaving their machines, the aircrew congregated beside a mobile FM radio installation, placed near the centre of the field. "There's been a cock-up," said the radio operator. "The army aren't ready for us yet. We've got to wait here for 30 minutes." We heard the sound of other Wessex helicopters approaching the field. In turn, they landed and closed down. Soon, there were over a dozen machines lined up in the field. The aircrew joked and chatted as they waited. Some of them smoked, and tramped the ground to help keep warm. I noticed that by chance a neat semi-circle had been formed.

Suddenly, I became aware of the noise of bag-pipes. I was standing quite near Frank as the familiar drawl of the pipes sounded.

"Remember what you said last night about the weird and wonderful world of helicopters?" I reminded Frank.

"No. Did I say that?" he asked.

"It's just a detail," I said. "But I can't help noticing that here I am standing in a muddy field, in a foreign country, nowhere in particular, doing not a lot, listening to bag-pipes."

"Slightly eccentric, I suppose," he laughed. "I'm sure there's worse to come."

We continued to stand around chatting until the half-hour delay had elapsed, then the aircrew returned to their machines. After starting, the helicopters lifted in sequence, and followed the leader to an area at the edge of some extensive woods. As we approached the woods, we spotted a number of military vehicles. Crouching near the vehicles were long lines of troops.

The lead Wessex was marshalled to a landing site, and the other helicopters landed in a row alongside the leader. Quite quickly, the troops started to run towards their allocated machines. The soldiers kept their heads low as they approached the turning rotor blades. They carried rifles, and heavy rucksacks. One of them glanced up at me as he waited his turn to board the helicopter. I noticed that he wore glasses; his pale face looked tired and strained.

To one side of the Wessex aircraft was a smaller Gazelle helicopter, waiting in a low hover. When the last Wessex had been loaded with troops, the crewman gave a 'thumbs-up' sign to the Gazelle. The Gazelle then slowly lifted, and started to head towards the border area between East and West Germany. In turn, all the Wessex aircraft lifted, and followed the Gazelle. The whole formation was strung out in a lengthy line-astern pattern.

"We're getting really close to the border now," said Frank as he studied our progress on his map. "It can be hard to see the actual border on occasions, with just small white posts marking the boundary. Sometimes the white posts are even hidden in undergrowth. Our Border Guard personnel know their patch of border really well, but still have problems occasionally. Over on the eastern side are the wire fences, look-out positions, dog runs, and mined areas. Let's hope we don't see those: it would mean we'd crossed the border."

Suddenly, we noticed that the Gazelle had turned sharply.

"Better watch out," said Frank. "He's spotted the border a bit late. Some of the East German Guards have been known to be quite trigger happy."

The Wessex helicopters followed the Gazelle's hard turn like a 'Mexican wave'. Soon we noticed that the leader's airspeed was reducing.

"Should be landing at any moment," said Dennis. "They must have chosen a spot close to the border to show a presence." Ahead we saw a sizeable clear area, and a marshaller wearing a bright day-glow jacket. The Gazelle maintained a low hover as the first Wessex landed near the marshaller. The other helicopters landed in turn. The troops disgorged rapidly from the aircraft. As soon as the last soldier had left, the Wessex helicopters lifted in order and hastily re-traced their flight path, following the Gazelle.

"The Gazelle will take us back close to the border again," said Frank. "Just to make our presence known. Then we'll return to the pick-up point and wait for instructions."

The convoys of military vehicles were still parked as we approached the pick-up point. The marshaller signalled the aircraft to close down once we had landed. The aircrew then climbed out of their machines, and walked towards the vehicles. They mingled with army Officers who were briefing, and studying maps. There was a general hubbub.

Suddenly, a familiar voice called me from behind.

"Richard!" I turned round. "Fancy seeing you here."

It was my cousin, Hew.

"Well, well," he said. "Talk about a small world."

"I know. Haven't seen you for ages. How's life?"

"Not bad. Not bad." We continued chatting. Slight in build, and with a shock of brown hair, Hew could seem an unlikely army officer at times. However, his droll voice and urbane surface belied a powerful character. As the Commander of 3 Parachute Regiment in the 1982 Falkland's War, he was credited with a backbone of 'pure steel'. His piercing eyes tended to dart about, taking in the surrounding scene. Some years in the future, as a Lieutenant General, Hew featured in the TV series 'Back to the Floor'. I discovered later that Hew and the White Tornado did not get along; the two strong personalities tended to clash.

As we talked, we suddenly heard the bag-pipes playing again.

Hew looked at me in amusement. "I guess," he said, "if we're going to have a family reunion, this place is as bizarre as anywhere!"

We looked about us, laughing at the comedy of the situation, as the sound of the bag-pipes continued to wail in the background.

CHAPTER 35

A Couple of Crises

The Operations Officer handed me the telephone receiver. My wife had called: "Could you come home as soon as possible?" she asked.

As I put down the telephone receiver, the White Tornado looked at me.

"Trouble?" he asked.

"Lizzie's not been well for a while now," I said. "It's rare for Sue to ring me like that."

"I know," replied the White Tornado. "You push off home pronto, my lad. There's not much on the programme today. What there is amounts to crap anyway, if you ask me."

As I left the Operations Room, the White Tornado called after me: "Good luck old boy," he said.

Hastily, I went to my car.

During the drive home, I mulled things over in my mind. Six months had gone by since exercise 'Gryphons Galore'; it was now early 1979. My family had joined me in the late Autumn of the previous year, to live in a small flat in Harsewinkel until a proper Married Quarter had been allocated. In the unhealthy 'muck from the Ruhr' climate, Lizzie had gone from bad to worse with a series of illnesses. The child was thin and wan; her skinny legs and arms had become stick-like. Her parents were upset and worried.

I noticed that the front door was cracked open as I bounded up the stairs leading to our flat. I heard the sound of voices as I went towards Lizzie's bedroom. Inside were two people: her mother, and Rosemary our neighbour, who was an ex-nurse. Rosemary was just taking a thermometer out of the child's mouth. Her expression was impassive as she inspected the instrument. Without fuss, however, she indicated to us to leave the room for a moment. We moved into the dining room area.

"I don't want to alarm you," said Rosemary quietly. "But you should call an ambulance immediately. Lizzie's temperature is over 104."

Quickly, I went to the telephone and made a call to the Medical Centre at Gütersloh.

"We'll be right over," said the Duty Nurse. "Pack an over-night bag for the child and her mother. We'll take them to the Military Hospital at Rinteln."

"Both of you go with Lizzie," Rosemary said to Sue and to me. "I'll look after young Alan."

The ambulance soon arrived, and Lizzie was stretchered into the back. The wail of the siren helped us to by-pass traffic as we made our way on to the *autobahn*. Sue and I looked at each other as the ambulance attendant tried to make her young patient comfortable. There was something surreal about the drive; the last time we had been involved in this sort of journey was for the birth of the children. Now one of them seemed in grave danger.

As the ambulance arrived at Rinteln Military Hospital, two nursing attendants hurried their young patient away with her mother in tow. I was asked to wait in the children's ward reception area. As I sat waiting, it was hard to concentrate on reading the magazines. Occasionally, I stood up and paced around the room. It was early evening, and the Hospital seemed quiet. I spotted a few other young patients in the children's ward.

At length, an army Doctor appeared. He seemed sympathetic and friendly; he clutched a clip-board. "We're still running tests," said the Doctor in a matter-of-fact way. "But I think she's got pneumonia. It's as well you brought her in when you did. We're working on getting her temperature down. You can go and see her now. I have suggested that her Mum stays the night here. We can put in an extra bed next to Lizzie's."

When I walked into her side-ward, Lizzie was lying on a bed with a fan blowing cool air on to her. She looked up, but found it hard to smile. Our young daughter seemed limp and sickly. The nursing staff were arranging the extra bed for their patient's Mum. We tried to comfort Lizzie, but she felt too ill to respond. She lay lethargically on the Hospital bed. She looked pale and debilitated.

After a while, a nurse approached me: "The ambulance is about to return to Gütersloh," she said. "They can give you a lift."

"You go," Sue said to me. "It'll be better for Alan. I can hold the fort here."

There was heavy traffic for the drive back along the *autobahn* that evening. I chatted occasionally with the driver and the nursing assistant. They were understanding and kindly. "It's an unhealthy place," said the nurse. "We have so many problems, especially at this time of year."

When I eventually reached our flat, Rosemary handed me a message from the White Tornado. I opened a small scrap of paper. A note had been hand-written: 'Sorry about Lizzie,' read the message. 'Stay at home as long as you need. Tomorrow's programme looks as crap as today's.'

❖ ❖ ❖

It was several days before Lizzie and Sue were allowed home. It took much longer for Lizzie to recover fully.

By the Spring of that year, the White Tornado seemed to be notably more up-beat about the programme: some unusual tasks had been planned. "Three aircraft have been invited to go to Aalborg, in Denmark," he said to me one day. "They'll be working with the *Jagercorps*, the Danish equivalent of the SAS. If you'd like to go, it should be an interesting trip."

"Yes please," I volunteered.

When the due date was reached, the three Wessex departed together from Gütersloh. The aircraft flew towards Denmark in a loose form of 'Battle' formation. Our flight-path took us north towards *Schleswig Holstein*. We passed by the urban sprawl of Hamburg, with the seaport of Lübeck to the east. Then we followed the *autobahn* on the northwards flight, crossing the famous Kiel Canal before approaching the border. Once in Denmark, we continued to follow the motorway, until it branched towards Arjus.

At that point, we were flying over high ground in the Jutland area of Denmark. Eventually, the town of Aalborg was sighted, with its airfield to the north-west. We spoke to Air Traffic Control: "Welcome to Aalborg," they said. "You are cleared to land at our airfield."

"Welcome," the *Jagercorps* Major also greeted us as we climbed out of the Wessex cockpits. "There's no flying planned for the rest of the day. Come and meet our personnel."

The distinctive smell of strong coffee wafted from the *Jagercorps* crewroom. Inside the room, a throng of folk stood around chatting. They glanced up, and greeted us warmly as the aircrew entered. There were looks of recognition. "Ah yes. I remember you from last time. Good to see you again." The 'old hands' were soon engaged in animated conversation. I was introduced as the 'new pilot'. It was striking to note the high esteem in which we were held by the *Jagercorps*. "We've put in special requests to work with you guys," said the Major. "We've had to wait our turn; it seems you have a busy programme. We're honoured to have you for a whole week."

The following day, we gathered early for briefings. The morning would be spent teaching the *Jagercorps* soldiers about parachuting techniques, which would be put to practise in the afternoon. 'Taff', my crewman, talked to the small squad of *Jagercorps* men allocated to our aircraft.

"Now then, gentlemen," said Taff in his distinctive Welsh lilt. "This can be extremely hairy unless you follow the correct procedures. We will have all the sharp objects around the cabin door covered with tape for your protection. The one thing we can't do anything about is the tail rotor. You must allow due time, therefore, between jumping from the cabin, and pulling your rip-cords." The soldiers carefully examined the parked Wessex. They familiarised themselves with the cabin area from where they would jump.

Eventually, Taff said: "If there are no more questions, gentlemen, we'll see you this afternoon." The soldiers then disappeared for their own discussions before the afternoon's flying.

However, the *Jagercorps* returned early that afternoon, eager to start. The soldiers wore life-jackets underneath their parachute harnesses. By chance, I saw that the Major, who was flying in the aircraft next to ours, did not wear a life-jacket. I pointed it out to Taff. "He's got dispensation," said Taff. "There's some problem. Anyway, he's the boss. He doesn't need one!"

The soldiers strapped into the seats in the Wessex cabin area, and waited as we started the machines. I noticed that the men appeared to be quite tense. They checked and re-checked their equipment. The soldiers glanced at each other nervously. The bulky parachutes looked uncomfortable.

Soon, we had obtained permission to take-off. The Aalborg Air Traffic Control Officer gave us clearance to climb to 10,000 feet, our maximum permissible height without oxygen. We would fly in a loose echelon formation, and the helicopters would monitor each other during the proceedings.

As we approached 10,000 feet, the other two aircraft were slightly ahead of our machine, and to the right of us. We listened on the radio to Aalborg Air Traffic Control, and obtained final consent for the para-drop onto the airfield. After just a short delay, I spotted the first man jump from the lead Wessex. I saw him fall rapidly as he counted the briefed number of seconds. Then I saw his parachute billow open. In prompt sequence, the other *Jagercorps* men followed.

Once the first batch of parachutists were clear, the lead Wessex did a wide turn, maintaining height so as not to interfere with the subsequent aircraft. Then we saw the next group of men jumping from the second Wessex. Soon it would be our stint.

I glanced back at Taff, and gave him a 'thumbs up' signal. "OK Taff," I said on the inter-communication system. "They're clear to unstrap." The men knelt down as they lined up in their allocated order. The *Jagercorps* did final checks of their equipment. The men looked at me anxiously, knowing that it would be on my command that their jump commenced. I wanted to ensure the correct spacing between each 'wave', as had been briefed. I had started my stop-watch.

"Thirty seconds to go," I said to Taff.

"Roger," he replied. The weather that day was fine, with only small amounts of cloud. However, there was a fairly strong wind from the north. Below us stretched the distinctive *fjord* to the west of Aalborg. I could see clearly the outline of the coast on both sides of the northern tip of Denmark. Expanses of sandy beach followed the coastline.

"Fifteen seconds to go," I looked at Taff briefly. He gave a 'thumbs up' sign. Underneath our aircraft, I followed the progress of the khaki-coloured parachutes from the first two Wessex. The parachutists descended slowly towards the airfield.

"Ten seconds," I called. At that point, I noticed that some of the parachutes below had started to drift further south than intended. The other two helicopters maintained a sweeping turn, keeping well clear of us.

"Five seconds to go," I said. The first man to jump was positioned by the cabin door. He held out both arms, gripping the sides of the door. Taff was just behind him, with one hand holding the soldier's shoulder.

"Four seconds..." I continued the count-down.

"Three seconds..."

The *Jagercorps* soldier made a final glance in my direction.

"Two seconds..."

The parachutist concentrated on looking outside now.

"One second... *You're clear to jump!*" I called.

Immediately, Taff released his grip on the first parachutist's shoulder. The man leant forward slightly, and then, in a deliberate technique, he leapt out of the aircraft. Smartly, the next in line took the first man's place. Taff tapped his shoulder, and he too leapt from the helicopter. Soon, all our soldiers had jumped. I counted the parachutes as they opened below us.

Suddenly, there was a call on the radio from Air Traffic Control.

"Heli-drop leader, this is Aalborg."

"Go ahead," said Al, the Captain of the lead Wessex.

"Roger, we think some of the parachutists are heading for the sea."

"Copied," replied Al. "Request clearance for rapid descent."

"You are so cleared," responded the Air Traffic Control Officer.

"The other two helicopters will follow me down," said Al to the Air Traffic Control Officer.

The second Wessex, flown by 'Buckloader', applied a high angle of bank as it tailed the lead machine. I followed both aircraft closely. Al made a wide orbit as he took his aircraft clear of the descending parachutes.

"I understand the problem now," said Taff. "This northerly wind is stronger than they anticipated. I can see that two or three of them are likely to end up in the sea. It's a good job they're wearing life jackets." Taff and I looked at each other. We had remembered at the same moment: there was one man without a life-jacket.

Soon, all three helicopters were descending through 1,000 feet. We kept well clear of the parachutists as we crossed the coast, heading out to sea. The lead Wessex then descended further to 500 feet. The other two machines followed, and we all flew an orbital pattern over the sea, just beyond the airfield boundary. We continued to watch the descending parachutes.

"From the Air Traffic Control Tower, it now looks like just two of them will end up in the sea," said the Air Traffic Control Officer.

"Roger," replied Al.

The two affected parachutists were approaching 1,000 feet at that stage. The three helicopters maintained a turn, keeping a safe distance from the descending parachutes.

Eventually, Al said: "I'll take the westerly one. 'Buckloader': you rescue the other one."

"Copied," responded 'Buckloader'.

"Dick, you monitor proceedings from overhead," said Al.

"Will do," I replied.

Both parachutists had descended towards 500 feet at that stage; one man was slightly lower than his companion. The other two Wessex had turned towards their respective rescues and loitered nearby, observing the final part of the soldiers' descent.

Suddenly, I noticed a splash as the first parachutist hit the surface of the sea. Al flew towards the man; the helicopter was flown into the wind, on a northerly heading. The down wash from his machine was thus blown away from the parachutist by the prevailing wind. Otherwise, there was a danger of down-wash interference as the *Jagercorps* man strived to detach himself from his parachute harness. I flew in a large orbit, observing both rescues happening beneath.

"This one's got a problem," Al said on the radio after a while.

Taff and I looked at each other.

"That's sod's law for you," said Taff. We both felt that we knew it was the life-jacketless Major who was struggling.

"He's being dragged under by his parachute!" The next radio call from Al revealed an excited edge to his voice. "'Buckloader'; how are you getting on?" he asked.

"Fine. No problem," replied 'Buckloader'. "This one's wearing a life-jacket."

"Roger," said Al. "I'm going to have to do something about this guy. I think he's drowning."

Taff and I peered down at the scene below. We noticed that the Wessex had moved closer to the parachutist than would have been normal. Then we saw the machine descend to the surface of the water: it was flying so low that the tips of the main-wheels were submerged. At that point, the crewman moved towards the small step just outside the aircraft cabin. He sat on the step; he had loosened his safety harness. The crewman's feet dangled in the water. The Wessex crept even closer towards the drowning man. Shortly, we saw the crewman lean forward. He was attempting to physically lift the parachutist out of the sea.

"He'll never do it," said Taff. "The weight of that saturated parachute will pull both of them down."

The crewman below then seemed to be moving his arms. "He's cutting away the parachute material," said Taff.

Time seemed to slow down as Taff and I continued to look anxiously at the drama, realising just how much the *Jagercorps* man's life was in jeopardy. The crewman himself would quickly tire under the circumstances; the situation was full of danger for him too. At length, after further cutting efforts, we saw the crewman haul the parachutist towards him. The soldier was being dragged onto the cabin door-step. Eventually, the soldier appeared to put out one hand to grip the step. The Wessex maintained its low hover; the door-step was kept just on the sea's surface. We imagined the soldier to be exhausted as he clung on to the helicopter. I maintained a low, slow orbit over the area, and I sensed that Taff felt frustrated at being unable to 'beam down' to help his colleague below. Then, at last, we noticed that the gap between the surface of the sea and the step was slowly widening.

"I think they've got him," said Taff. "They're slowly lifting him clear of the water."

The parachutist and his attendant crewman seemed to be pulling themselves gradually into the aircraft cabin. The helicopter maintained its steady hover over the sea as the men struggled. At length, both the men disappeared inside the cabin of the Wessex, and we saw the cabin door close.

"Aalborg, this is the Heli-drop leader," we heard Al's call on the radio. There was a note of relief as he spoke.

"Go ahead," replied the Air Traffic Control Officer.

"Roger. Rescue now complete. Request clearance direct to the Medical Centre."

"You are clear," said the Air Traffic Control Officer. "The Medical Centre has been alerted."

Taff and I continued to monitor the progress of the rescue Wessex. 'Buckloader' had completed the recovery of his *Jagercorps* parachutist by that stage, and was shutting down his machine in the dispersal area. Once the Major had been delivered to the Medical Centre, Al and I returned to the dispersal area as well.

After closing down the helicopters, the crews teamed up together, and discussed the dramatic events. It was decided to cancel further parachute drops for that day.

However, later that evening, we had a briefing in the *Jagercorps* crew-room about the activities planned for the following day. The room was packed with people. Maps were studied; an over-head slide projector was in use.

Suddenly, in the middle of the briefing, the door at the back of the room opened quietly. Everyone turned round to make out who had entered. The *Jagercorps* Major seemed apprehensive as he poked his head round the door. The folk in the room immediately recognised him. Spontaneously, everybody present broke into applause.

The Major beamed: "I want to thank you guys," he said eventually. "You saved my life today." He appeared to be uncharacteristically emotional as he spoke.

Helicopters returning from a field exercise, Gutersloh 1975.

CHAPTER 36

Trouble in Tin City

"We have an extra-early start tomorrow morning," said the *Jagercorps* Major. He had recovered from his parachuting incident, and was briefing the planned activities for the last day of our detachment in Denmark.

The three Wessex were required to be airborne before dawn. The machines would fly along the *fjord* due west of Aalborg, to pick up teams of *Jagercorps* soldiers just before first light.

Fortunately, when we took-off the following morning, the weather was fine. A half-moon was hanging in the sky, and the stars were clear in the cold pre-dawn air. The reflection of the moon on the surface of the *fjord* shimmered eerily as we flew away from Aalborg.

The three Wessex followed each other in line astern formation. We hugged the coastline, knowing that the area was clear of masts and other high obstructions.

"I'm knackered," complained Taff. "This is a crazy hour to be flying. As for flying without lights…" We had been briefed to turn off our navigation lights for tactical concealment when close to the pick-up point.

Soon, we approached the planned troop recovery area. The lead Wessex switched off its navigation lights. 'Buckloader' in the second helicopter followed suit, and I switched off our lights. The crews listened on a pre-briefed frequency, but did not speak unless there was an emergency.

The outline of the leading helicopter was faintly visible in the moonlight when it turned towards its designated field. 'Buckloader' then changed course as he headed towards his nominated landing spot. I started my stop watch.

"Turning in ten seconds," I told Taff.

"Roger," he replied.

I applied bank at the due moment, and at the same time began to reduce our airspeed. As we pointed towards our intended pick-up point, Taff opened the Wessex door so that the pair of us could search for the agreed signal: intermittent flashes from a green-lensed torch.

"Can't see anything yet," said Taff.

We continued to search. As we had no lights, the troops relied on the noise of the helicopter to alert them. Suddenly, ahead and to our right, we recognised some green torch flashes.

"That's them," said Taff. "We should make out the shape of the 'T' soon." The soldiers had been briefed to hold torches in the shape of a 'T'; by flying towards the cross-bar of the 'T', we would be pointing into the wind at the landing site. As we got closer to the men, the 'T' shape magically appeared.

"We're slightly out of wind on this heading," I said to Taff. "I'll start making heading corrections now." I was flying slowly at that stage; the heading corrections soon put us into the wind.

As we continued to make our stealthy approach, Taff said: "It's too dark to see properly, but from what I can make out, the area looks clear." The landing site would have been carefully checked out for clearance from obstructions the day before. Then the helicopters would be allowed to make a safe 'lights out' approach.

"That's fine," I replied. "They seem to have their landing 'T' established pretty much into the wind. I'm sure the area safety check yesterday was done in their usual thorough fashion."

We soon got closer to the landing 'T', and I noticed that the bottom part of the 'T' had started waving about under the influence of the helicopter's down-wash. "That man nearest to us looks a bit wonky," I said to Taff.

"It must be quite frightening standing there in the pitch dark as this noisy brute arrives seemingly from nowhere, and starts blowing them around," commented Taff.

At that moment, the bottom part of the 'T' suddenly disappeared.

"He must have taken fright," said Taff. "He's probably hiding behind a bush."

I continued to make the slow, steady approach, relying on the shape of what remained of the 'T' for guidance. Suddenly, the 'T' changed shape again.

"That's another one pushed off," said Taff. "Soon there won't be anyone left."

As if my crewman's words had been over-heard, at that moment the entire right hand section of the landing 'T' disappeared. I was left with a straight line of just three green torches.

"This isn't so good," I said to Taff. "I'm reluctant to turn on the landing lights at this stage. It'll destroy our tactical concealment in an instant. It'll also wreck our night vision, and the soldiers'. But I'm not happy about landing without more guidance. I just can't see well enough." At that point, two more members of the 'T' vanished. I was left with just one brave soul holding his torch.

"Standby," said Taff. "I'll point my torch at the surface, and see if I can talk you down." At that stage I had brought the Wessex to the minimum safe airspeed. Coming to a high hover was not an option. It was too dark to see sufficient outside-cockpit visual references for hovering, and maintaining a hover on instruments was not possible: the instruments were insufficiently accurate.

"OK," I said. "If it doesn't work, I'll have to overshoot."

"Roger," said Taff. He was crouching by the open door-way, as he pointed his night flying torch below the helicopter. "I can just make out the surface," said Taff. "Maintain this airspeed, and descend slowly. Confirm you are still visual with the last member of the 'T'?"

"Affirmative," I replied. "He seems to be holding his ground."

"Good man," said Taff. "Continue forward flight; estimate range ten." I maintained the slow forward flight, and gradually reduced our height.

"Continue descent," said Taff. "Further forward a range of five." Fortunately, the remaining stalwart held his torch firmly as we crept closer to the landing site.

"Forward range four," called Taff. We approached a dark spot; in spite of the ambient moon-light, my main visual connection with *terra firma* was from the torch-light ahead.

Then Taff said: "Forward range three. No lower." I made a slight upwards movement with the collective lever to prevent further descent.

"Forward range two," commanded Taff. I still relied on the remaining member of the 'T'.

"Forward range one," said Taff. Then he called: "Steady. Steady." At that stage, I was just able to make out the faint surface glow provided by Taff's night flying torch. With the combination of that and the remaining member of the 'T', I had just sufficient visual references to maintain a hover.

"Confirm able to continue a hover?" Taff asked me.

"Affirmative," I replied.

"Well done," said Taff. "In that case down five." Using the tenuous references, I managed to sustain a hover as I eased down the collective lever. Taff continued the count-down: "Down four…down three…down two…down one…just touching." At that point, I felt the wheels make light contact with the surface of the field.

"From what I can see with this torch," said Taff. "It looks OK. The wheels are holding." I fully lowered the collective lever; the machine rested onto its oleo struts.

"That still looks fine," said Taff. "The wheels aren't sinking."

"Roger," I replied. "In that case you're clear to make the signal for boarding." Taff made a series of intermittent dashes with his torch. Within moments, a group of soldiers ran towards the cabin door. Their torches flashed around them as the men hastened to the helicopter. Taff guided them to their seats.

"They're all strapped in," said Taff eventually. "You are clear to lift." I checked my watch. We were in the allocated time band to allow us to get airborne with safe clearance from the other two Wessex. I was thus able to take-off without having to make a radio call. If outside the time bands, we had briefed to make a safety call to establish each other's position.

I pulled up the collective lever, and made a 'towering' departure as rapidly as possible. Then I followed the pre-briefed safe route back to the edge of the *fjord*. Once clear of the tactical landing site, we had arranged to put on the aircraft lights again. I soon recognised the lights of the other two helicopters several miles ahead.

As we re-traced our route along the edge of the *fjord*, the first signs of dawn were just appearing. A pale light grew on the horizon; the moon and stars also remained visible at that stage.

"Aalborg, this is the heli-leader," we heard Al's voice as he called the Air Traffic Control at Aalborg.

"Go ahead, heli-leader," replied Aalborg.

"Roger, approaching ten miles from your airfield," continued Al. "Request visual recovery. There are two helicopters following me."

"Copied heli-leader," said Air Traffic Control. "No conflicting traffic. You are clear for visual approach."

The bright lights of Aalborg airfield made a startling contrast to our 'lights out' pick-up. At length, all three Wessex landed and shut-down in the dispersal area. We were met by the Duty Operations Officer. "Time for breakfast, you guys," he said. "I think you've earned it!"

The aircrew sat grouped together during their breakfast, and on adjacent tables were the *Jagercorps* troops we had just picked up. When the breakfast was nearly over, a *Jagercorps* Sergeant came up to our table.

"Excuse me gentlemen," he said in his accented English. "May I ask who were the crew of the last machine?" Taff and I nodded at the Sergeant.

"I must apologise," continued the Sergeant. "I'm afraid it was rather overwhelming when the helicopter got close to us in the dark conditions. The men were not happy!"

"Who was the one that stayed?" I asked.

"It was me," replied the Sergeant.

"You saved the day," I said.

"Don't you mean night?" interjected Taff.

The Sergeant beamed.

We walked back to the *Jagercorps* crew-room after breakfast. Plans for the day were being prepared, including training the soldiers in abseiling from the helicopters. Suddenly, the *Jagercorps* Major appeared.

"Gentlemen," said the Major. "I have had an urgent message from your base. Exercise 'Peg-out' has just been called. You are required to return to Gütersloh immediately."

The aircrew greeted the news with dismay. We had looked forward to the day. The flying programme was interesting, and a farewell party had been planned for the evening.

"Oh well, I guess that's that," said Al eventually. "What a pain in the arse. Still, it can't be helped. I suppose we'd better go and pack our kit."

"We could always make out that we got the wrong message," suggested someone.

"Or that the aircraft all broke down," was another offering.

Eventually, it was decided to bow to the inevitable. By late morning, the crews had shaken hands with their *Jagercorps* hosts, and pledged keenness to return at the earliest opportunity. The aircraft then took-off from Aalborg, and headed towards the airfield at Schleswig in northern Germany, where a refuel had been planned. The transit flight continued towards Gütersloh after the refuel. Eventually, we heard Gütersloh Air Traffic Control on the radio.

"We have a message for you," said the Air Traffic Control Officer.

"Go ahead," replied Al.

"Roger. You are to proceed directly to field locations. Standby for separate map references for each aircraft."

Taff and I looked at each other. "That's a bit rough," said Taff. "They're not even allowing us to go home to freshen up, and get clean clothing and stuff."

Al turned the formation towards the field sites we had been allocated, and at length said on the radio: "Thanks for your help everyone. See you at the end of 'Peg-out'." The three Wessex then went in separate directions to their different field locations.

When Taff and I landed at our allotted site, it was early evening. We were met by Pete, the Flight Commander of 'D' Flight. "Sorry about this," said Pete. "We've got some heavy tasking lined up tomorrow, and we're short of aircraft. It'll be an early start. The good news is that we're located on a particularly commendable farm: positively five-star."

Taff and I unloaded our kit from the Wessex, and Pete pointed to a barn. "Make your way over there guys," he said. "We're using the upstairs part." Fresh straw had been strewn over the barn's floor. Taff and I chose our pitch, and placed sleeping bags there.

"Better go and get a bite," said Taff. "Then it'll be time for some kip. I'm still knackered from this morning!" It seemed a world away from our dawn pick-up of *Jagercorps* troops.

❖ ❖ ❖

"This is going to seem quite weird," Pete had said the following morning. "You two have been tasked to go to 'Tin City'." Taff and I noted Pete's puzzled look. "This isn't what they planned originally. But apparently there's some trouble." Pete had no further details, but we were required there as quickly as possible.

As we flew towards 'Tin City', Taff asked me if I had been there before. "No," I replied. "But I've heard about it. It's where the army teach soldiers about riot control techniques."

Standing alone in open countryside, 'Tin City' could be seen from some miles away as we approached. It was a haphazard arrangement of buildings, surrounded by a wall of corrugated iron.

As we landed the Wessex, and shut-down the machine, a figure hurried out of a side building. An army Captain approached us

"Gentlemen, please follow me," said the Captain tersely. Taff and I were taken into the small building the Officer had just left.

"I hope you've got your gas-masks," said the Officer as we entered the building.

"We're on field exercise," I replied. "We carry them at all times."

"Good," he said. "You may need them today." Inside the building, there were more Officers. They were having discussions as they pointed to a detailed map of the site.

"The thing is," the Captain said to Taff and to me. "Matters sometimes get a bit out-of-hand here. We try to make the training as realistic as possible. The two 'sides' have been known to get carried away. If it goes too far, and there's danger of injury, we have to call in the Military Police. The 'Red Caps' then go in for real with tear-gas. It doesn't happen often, but I'll be frank: it's looking quite dodgy at the moment. We may need you to fly in the Military Police."

As the Captain spoke, Taff and I were aware of a considerable clamour coming from the direction of 'Tin City'.

❖ ❖ ❖

Mitrovica, Kosovo, February 2000. Twenty-one years in the future.

Tens of thousands of ethnic Albanians had set out one week-end from Pristina, the Capital of Kosovo, on a twenty-five mile march to the northern town of Mitrovica. The swell of people had been met by United States military vehicles and soldiers lining the road leading to the centre of Mitrovica, to prevent them from entering. However, thousands of demonstrators broke through to reach a bridge where another group of local Albanians had gathered.

French troops, who were permanently stationed in Mitrovica, had been criticised by UN police and local leaders for failing to provide adequate security.

British troops from the Royal Green Jackets were eventually called in by the French. In the dangerous situation, the French had considered that the riot control training given to the British troops would help them to manage the problem.

The Royal Green Jackets had elected not to wear the intimidating-looking gear of the French riot troops. The British soldiers had worn berets; their Colonel had correctly assessed that the Kosovans would be calmed by a low-key approach. He stood on a vehicle, and with the help of an interpreter, told the crowds that the troops were there to help. In essence they were on the same side; they were not enemies. His tactics worked. Although the jostling continued, it became less hostile. The crowds were placated by the British approach. The ethnic hatred was not resolved; in the longer term there was more violence to come, but at least on that specific occasion the thorough training had paid dividends.

❖ ❖ ❖

May, 1979. The noise from 'Tin City' continued; an atmosphere of commotion was created as dustbin lids, and other metal objects were struck with wooden batons. The Officers in the Operations Room pursued their discussions. Heads nodded anxiously as the situation was debated. The Officers had to encourage the soldiers to gain valuable training experience, without allowing the matter to get out-of-control. It was a fine line which had to be negotiated.

Suddenly, one of them approached me. "It's simply an idea," he said. "But sometimes just the sight and sound of a helicopter has an effect. Before we call in the Military Police, could you fly slowly around the compound?"

"No problem," I said. "Would you like to come with us?"

"Sure," replied the Captain.

Three of us hastened to the Wessex, and I ran through the start-up procedure as Taff monitored the start from outside. Soon, Taff returned to the cabin, and I lifted the machine into a hover once he had strapped in.

"Clear above and behind," called Taff. I raised the collective lever, and the helicopter rose vertically. As we climbed up in a 'towering' type of take-off, I maintained our position over the same spot on the ground. We soon became level with the top of the metal wall surrounding 'Tin City', and we were then able to look into the compound. Men could be seen running around. Several carried sticks, batons, and other objects. A few looked up at the helicopter as it rose slowly above their heads.

We had strapped the Captain into the spare cockpit seat to act as observer. "That's handy," he said. "We get an over-view of what's going on now."

The Captain then proceeded to outline what was happening. "From the operations building, we have radio contact with the leader of the 'riot control team'," said the Captain.

I maintained our position in a high hover as the Captain spoke. "The 'riot team'," continued the Captain, "were pre-positioned to create a 'riot'. The 'control team' were then sent in. Having subdued the 'riot', the 'control team' are supposed to withdraw. However, this group of 'rioters' are a lively bunch. They're not allowing the 'control team' to withdraw and re-group."

THE SPICE OF FLIGHT • 241

We continued to watch proceedings within the compound. Then the Captain requested that I flew a slow circuit around the compound. We noticed heads turn to observe us; our presence seemed to provide a distraction. After a while, we saw a group of men collect near the entrance area.

"I think we should land now," said the Captain, after we had completed the slow flight around the compound. We returned to our original landing site, and the Captain hurried off as Taff and I went through the shut-down procedure. We then walked to the operations building. Inside, there were still discussions in progress, but the Captain seemed to have a look of relief. He came towards us.

"It worked," he said. "As we thought, the presence of the helicopter seemed to have a sobering effect. The 'riot control team' are leaving shortly, to re-group outside the compound. There'll then be a 'cooling off' period, and they'll try again. It's all been valuable training. The soldiers will learn that the 'heavy-hand' approach isn't always the best way to deal with some situations. However, we'd like you to remain on standby here in case of further trouble."

In the event, there was no further trouble, and by the end of the day Taff and I were released from our watch over 'Tin City'.

We returned to the field location, and Taff and I continued with 'D' Flight for the next ten days of exercise 'Peg-out'. To Pete's regret, after two days in the 'five star' barn, we were moved on. We attached ourselves to another farm for the next few nights. "This one's only three star, I reckon," commented Pete ruefully.

Eventually, 'endex' (end of exercise) was called, and the White Tornado organised an in-flight rendezvous of all the Squadron aircraft. An impressive return of sixteen Wessex helicopters, in groups of four, was witnessed by folk at Gütersloh.

"At last," Taff had said to me as we unpacked kit from our aircraft. "It's about time these shirts and shreddies had a good wash."

With the combination of Exercise 'Peg-out', and the detachment to Denmark, we had not seen our families for nearly three weeks.

By that stage, Sue and I had moved from our flat in Harsewinkel. We lived in a Married Quarter on 'Top Patch'. We had a house instead of a flat, and the family seemed happier in the new environment. All the houses had cellars, and in our cellar, we had set up a washing machine and other laundry arrangements.

My welcome home that day was typically effusive. Up until the time of the telephone call, the family were chatting and bantering – happy to be together again. When the call did come, I was downstairs in the cellar, loading the washing machine with the three weeks' worth of unsavoury 'shirts and shreddies'. The telephone was answered by Sue. I could not hear her conversation, but after a moment, Lizzie appeared. She looked anxious, and pointed upstairs.

As I climbed the stairs, Sue was finishing her conversation. She replaced the telephone receiver just as I reached the top step. Sue continued to stare at the telephone. She had a look of disbelief. Her face appeared white and drawn.

Eventually, she came up to me and put her arms around my neck.

I did not hurry her. The news would come out in her own time. At length, however, I heard the story. I gathered the information that her mother had been admitted to hospital following a severe stroke; she was not expected to survive for more than 24 hours.

Lizzie and Alan gazed up in bewilderment as the tears streamed down their Mum's face.

Berlin, 1980

244 · THE SPICE OF FLIGHT

CHAPTER 37

Berlin – 1980

One year later. May 1980. "Oh look," said Sue. "See the Border Guards? And all that barbed-wire?" The train was just crossing the border between West and East Germany. Behind us was the Station at Helmstedt, in West Germany. Three miles ahead was the Station at Marienborn, in East Germany. Our four hour train journey, which had started at Braunschweig, would eventually end in the British Sector of West Berlin.

"I suppose it'll all come tumbling down one day," Sue continued to point at the barbed-wire.

I looked at my wife. Our third child was expected in November; she seemed vulnerable as she sat in the corner of the comfortable carriage. The year since the death of her mother had passed quickly. After the telephone call from her sister Jenny at the end of exercise 'Peg-out', Sue and I had hastened back to the United Kingdom; the military 'compassionate leave' system had swung into action with flawless efficiency. Even a taxi had been provided at Heathrow Airport to take us directly to the hospital. But Sue's father had recommended that we did not visit the hospital. Events had gone too far; the patient was in a deep comma, and there was nothing to be done. We had followed the advice, but regretted it later. We had missed the final opportunity for a personal farewell before Sue's mother died the following morning.

Now it was Spring-time one year on, and our time in Germany would be finished by the end of the year. We were keen to make the historic visit to Berlin before leaving Germany.

We had a first-class compartment on the military train almost exclusively to ourselves. The daily journey of the 'Berliner' train was,

THE SPICE OF FLIGHT • 245

in effect, a political act. It ensured that the right of rail access through East Germany was maintained. British military personnel were encouraged to make use of the train; the ride was free.

"That guy's peering hard at us," said Sue. The East German Guards were renowned for training their binoculars relentlessly on visitors. In addition to the barbed-wire, we saw the look-out towers, and the ploughed areas which denoted minefields.

"They'd come over to our side in an instant, given half-a-chance," I said.

The train was slowing as it approached Marienborn Station. The Station itself appeared deserted, apart from several officials in uniform. Once the train had stopped by the platform, three Officers in British military uniform stepped out. They marched stiffly towards a Soviet Officer and his entourage. The Soviet Officer then inspected the papers and documents handed to him. He took his time, and made some comments to the interpreter.

Meanwhile, we noticed a group of officials passing through the train, attaching wooden struts to the inside of all the doors. "It's to prevent anyone from East Germany trying to board the train," one of the officials pointed out.

"They must be desperate, poor things," said Sue.

When the formalities had been completed, the British Officers saluted the Soviet cortege, performed a ceremonial 'about turn', and marched back to the train. Meanwhile, we had picked up an East German train Guard.

Once the Officers had re-boarded the train, our journey to Berlin resumed. The stop at Marienborn had taken about half an hour.

When we were under way again, the differences between East and West Germany soon became evident. The farms on the east side of the border appeared to be managed as smaller units. The fields were minor in size, and tended by workers using antiquated, fragile-looking equipment. Even the occasional horse was seen pulling a plough. When our train crossed a road, there seemed little evidence of traffic. The cars we did see were Trabants, and other small Eastern-bloc makes. It was in stark contrast to the mass of large Mercedes and other cars which crowded the *autobahn* system in Western Germany.

After some thirty minutes or so beyond Marienborn, the twin spires of Magdeburg Cathedral came into view. By this stage the train attendant was serving us with afternoon tea. "Look over on the right side shortly," the attendant said. "You'll see a building they use for political prisoners." The train was travelling quite slowly as it passed through Magdeburg. The building mentioned by the attendant stood out austerely.

The train did not stop at Magdeburg, and having crossed the River Elbe beyond the City, our speed began to increase. At that point, our journey was around half-way through, and the attendant came up to ask if we wished to take supper in the dining car in a while. "It's highly recommended," he said. "And all part of the service. You may buy wine if you wish, otherwise there's no charge."

We eventually made our way towards the dining car, and were shown to a luxurious looking table for two. The linen was starched, and the table had been laid with impeccable care. The aura in the dining car was comfortable and old-world. We smiled at our fellow passengers, who seemed as impressed as we were.

As the attendant offered us the menu, he described the soup of the day. He then pointed outside, to a large engineering set-up. "That's the Soviet Tank Repair Workshop at Kirkmoser," he said.

The train continued to travel at speed while the appetising meal was served. Eventually, we felt the train begin to slow down. "We're just approaching Potsdam," said the attendant. "Our engine will be detached and searched there, and the East German train Guard will leave."

As we waited at Potsdam Station, it was an unusual opportunity to see some local people. There was a run-down atmosphere, and the presence of military personnel was noticeable. Folk seemed to move about lethargically; their clothing looked drab and colourless.

Once the formalities had been completed at Potsdam, the train started to move once more. Soon we spotted the 'Wall', look-out towers, and armed personnel which marked the border between East Germany and West Berlin. By that stage we had left the dining car, and the attendant came through the carriage informing passengers that the train would reach the West Berlin Station of Charlottenburg within ten minutes.

As the train drew in to Charlottenburg Station, we soon recognised a normal 'western' atmosphere. The Station was packed with people, and the citizens wore fashionable, attractive clothes. Newspaper vendors and shop kiosks surrounded the Station precinct. Smart advertisements were arranged on the palisade; taxis moved about busily. There was life and interest all around.

It was evening by then, and our taxi soon drew up outside the Edinburgh House Hotel, which was owned by the UK Government. The receptionist confirmed our booking, and we were shown to our room. It seemed a comfortable and well organised establishment.

The following morning, at breakfast, we chatted with the couple next to our table. During the conversation, they said: "We'll give you a word of advice. Get in touch with the Royal Air Force Police," urged the couple. "They make regular visits to East Berlin. It's a fascinating ride, and a rare opportunity."

After breakfast, I consequently made a telephone call. "We'll pick you up at 1000 hours," the Duty Policeman had told me. At the appointed hour, a white Volkswagen van with a 'Royal Air Force Police' sign drew up outside the hotel. I had been told that it was necessary for me to wear my Number One uniform.

As Sue and I climbed into the van, the police driver outlined the routine. "First of all," he said. "We'll drive to the observation post on this side of the Wall. After that, we'll cross through the famous 'Checkpoint Charlie'. Then we'll drive around East Berlin for a couple of hours."

At the observation post, there was a tragic scene of division. On the west side of the Wall, the facades of a once prosperous shopping street were all that remained. Behind the facades stood the rough concrete blocks which formed the Wall itself. Behind the Wall were the dog-runs, Watch Towers, search-lights, and mined areas which all obstructed the free movement of citizens. On the East side of the Wall, dismal blocks of flats over-looked the area.

"It was amazing what happened here on the 13th of August 1961," said our police guide. "Some people who went to Church were unable to get home after the Service. Massive rolls of barbed wire – the original Wall – went up rapidly and ruthlessly. Many folk were stuck on the wrong side. They were not allowed to cross the border, and a lot of families remained split-up."

We looked at the crosses and wreaths at the base of certain parts of the Wall. "Anyone attempting to escape from East to West is shot on sight," continued the policeman. "There have been some dramatic attempts. One teenager (Peter Fechter) was shot and injured when he tried to make a getaway. He lay for an hour or so, and West German police threw first-aid packets at him, risking their own lives. The 18-year-old kept calling out: 'Help me!' Eventually, the East German Guards carried him away. By then it was too late, and the lad died."

We walked slowly back to the van as the policeman told us more sad tales about the effect of the Wall. As we re-boarded the van, the policeman said: "We'll head for Check-point Charlie now."

A chicane of concrete blocks and other obstacles had been erected in the area of Check-point Charlie. "There have been several efforts by people to ram their way through here using lorries or other heavy vehicles," said the policeman. "The East German Guards know most of the tricks by now. As for us, they are obliged to accept our military passes and UK passports." We had parked the van, and walked to a small museum at Check-point Charlie. Inside the museum was a history of the Wall, and some graphic photographs of escape attempts.

After time spent in the museum, we returned to the van. The policeman then drove towards the border. We were waved through by the West German Guards, but as we approached the East German side, our driver slowed down and said: "They can be quite nasty at times, and cause deliberate delays. The best thing to do is look straight ahead, and try to remain expressionless."

Our driver slowed even more as we approached the East German Guards. There were a number of Guards, all armed, and we had the impression that they were checking each other as well as observing us; escape attempts by the Guards themselves had been known. We had been told to keep the vehicle windows closed, and the doors locked as we passed through the border. The driver held up his military Identification Pass against the side-window. I did the same against the window by me, and Sue held up her passport. The driver did not stop, but he drove very slowly past the East German Guards; they appeared to begrudgingly shuffle out of the van's way.

Once clear of the border area, the policeman took down his Pass, and turned onto a road which paralleled the Brandenburg Gate. On our right stood a tall radio mast, with a distinctive bulge in the middle. To the left, we could examine the apparatus which made up the Wall from the point of view of East Berliners.

"On this route," said the policeman, "we'll pass the bunker where Eva Braun and Hitler committed suicide on the 30th of April 1945. It was the day after their marriage."

As we drove past the undistinguished looking bunker, the policeman suddenly said: "Please don't turn round now, but we're being tailed by two *Stasi* secret police vehicles. This sometimes happens. If they're given an excuse, they'll stop us. I'll have to stick rigidly to the speed limit."

The driver said little as he concentrated on the journey. He was keen to avoid being held up by the *Stasi*. We were heading for a Soviet War Memorial at that stage, and as we pulled into the car park he said: "They're unlikely to bother us here." When we looked behind us, the secret police cars were no longer in evidence.

We stepped out of the van, and our police driver said: "I'll have to stay by the van, but you can walk around for twenty minutes or so if you wish."

In the park area of the Soviet War Memorial, various monuments had been erected. A long flight of steps led up to high walls which had been built on either side of the top step. At the base of these walls were statues of soldiers. The soldiers carried a rifle in one hand, a tin helmet in the other, and posed on bended knee. Above the soldiers' heads, painted onto the walls, was the sign of the Soviet 'hammer and sickle'.

Beyond the statues, another flight of steps led to a tall obelisk. An opening at the base allowed people to walk around inside. It was not crowded; just a few dozen folk were in evidence. Above the base, a monstrous effigy had been built. The obscure figures looked as sad as the whole Soviet regime, with its remaining life-span of just nine years.

Sue and I discussed the scene as we walked back to the van. "This place gives me the creeps," said Sue. "There's a complete absence of any sense of religious faith."

We approached the van at that stage. "No trouble during our absence, I trust," I asked the policeman.

"No, its been quite quiet, fingers crossed," he replied.

We drove off through the dreary streets of East Berlin. Not many people were seen, apart from a queue of passive-looking folk waiting outside a bread shop. In general, there was an absence of vitality and movement; a stark contrast to West Berlin. A few smart buildings stood in isolation: "They're the show-pieces to impress foreign visitors," said our guide.

As we approached a row of shops, the police driver asked us if we would like to look inside. "Yes please," said Sue.

"The official rate of exchange is one *Ostmark* (the East German currency) to one *Deutschmark*, (the West German currency)," said the policeman. "The rate of exchange in the real world is ten *Ostmarks* to one *Deutschmark*."

The police driver pulled up by a shop that sold glass. "You might find this one interesting," said the driver. Inside the shop, it seemed deserted apart from a woman shop-keeper. She appeared nervous and edgy as we looked around. The glass on display was limited, and the stock was apparently of poor quality.

After a few minutes, the shop-keeper beckoned us to the rear of the shop. She looked about her, and then pulled back a curtain. Behind the curtain was a door which led to another small room; inside were further displays of glass. "This is our more precious stock," said the shop-keeper.

"She's not wrong there," Sue whispered to me as we walked around the 'secret' room. "Some of this stuff is Dresden China – really valuable." Eventually, we chose some items of glass, and paid with *Deutschmarks*. "We've probably paid over the odds by East German standards," said Sue. "But these prices are still very low."

As we re-joined the police driver he said: "It's time to be heading back to the border now, but *en-route* we can watch the Guard Changing ceremony." The driver insisted on staying beside his van again, but Sue and I walked up to a building with tall columns at the front. The old-fashioned building housed an 'eternal flame'; it looked grand and formal. Groups of people stood in the area to watch as goose-stepping soldiers marched around in troops of three. "It looks most unnatural," said Sue.

We then went back to the van, and the driver headed towards Check-point Charlie for our return border crossing. "It'll be the same procedure as before," said the driver. He slowed to a crawling pace as we drew level with the East German Border Guards. The driver and I held up military passes; Sue pressed her passport against the window.

Suddenly, just as we passed the East German Guards, I noticed that the driver glanced anxiously in his rear-view mirror. He appeared to accelerate slightly. The driver headed determinedly through 'no man's land' for the safety of the West.

As we were waved through the West German part of Check-point Charlie into West Berlin, the driver sighed.

"What happened there?" I asked. The driver hesitated for a moment before replying.

"I'm not sure," he said eventually. "But one of those East German Guards seemed unhappy about something." We were securing our military passes at that point; Sue was returning her passport to her hand-bag.

"It was probably me," said Sue after a while. The policeman and I both looked at her.

Sue shrugged her shoulders as she said: "All I did was wink at him." The atmosphere was one of relief as all three of us burst out laughing. "World War 3 nearly started by a wink," commented our guide.

CHAPTER 38

A Funny Old Life

"It's a funny old life!" Mike was sentimental as he surveyed the unusual scene before our eyes.

We stood on the summit of a mountain in France. Just below us, a few hundred meters away, our Wessex helicopter was parked on a wide contour which provided an ideal parking place. Standing near us were a number of hill walkers who had taken several hours to reach the summit. The walk from the helicopter had taken us around fifteen minutes. "This is the way to do it!" Mike had remarked as we clambered up the peak.

We were on detached duty in France, sent from our normal base in Germany. Our task, and that of one other Wessex, was to provide training in helicopter support operations for a force of British soldiers. We were based at La Courtine in the mountainous territory of the Massive Central. On most days the training programme had been completed by noon.

"You can use the machines as you wish in the afternoons," the army co-ordinating Officer had said.

"You mean we have carte blanche?" we asked.

"I guess so," he replied. "They'll make good taxis while you explore the area."

"It's a 'jolly', really," the White Tornado had said. "But I suppose there's some training value." It was towards the end of 1980, and my tour with 18 Squadron was drawing to a close. The detachment at La Courtine had been offered almost as a 'swan song'. Mike had been scheduled as my crew-man; it was a rather different scene from our time together on the survival exercise in Bavaria.

"It's a funny old life indeed," I said to Mike eventually. "Anyway, time's getting on. We'd better return to the camp." There were looks of curiosity from the hill-walkers as Mike and I strode down the hill in our aircrew flying kit. As I climbed into the cockpit, Mike remained outside to supervise the start routine. At a judicious distance, several hill-walkers formed a group as they observed our activities. Some late-comers hurried up the hill to investigate the curious happenings.

Once 'turning and burning', I lifted the helicopter into a hover, and we returned the friendly waves of hill-walkers as we flew off towards La Courtine. We were in an area south of Clermont Ferrand, so I took up a north-westerly heading for the flight back to base. The helicopter made an ideal platform for our role as tourists; the route took us over spectacular terrain, a mixture of high ground interspersed with the occasional lake. Two days previously, we had spent time by one of these lagoons; a lake-side café owner had offered us lessons in wind-surfing.

Mike and I had been taken aback on more than one occasion at how the presence of the helicopter, with its British military aircrew, often had aroused strong reactions. The older generation had not forgotten World War Two, and the violent divisions caused by the Vichy regime. The town of Vichy itself, an old Roman health resort, was just north of our area. Many of the veterans with whom we came into contact were keen to show us evidence of their personal loyalty to General De Gaulle. It was thirty-five years since the end of the War, but their disgust and shame at the 'collaborators' still aroused strong emotions.

Approaching the Plateau de Millevaches we flew over a spectacular sight. "Look at that," said Mike. "What a fantastic view." Beneath us lay the deep ravine carved by the Dordogne River. To the east was the Haute-Loir and beyond that the Rhone valley. West of us, the high ground diminished towards the low-lying area surrounding Bordeaux.

As we came near to the army base at La Courtine, we suddenly heard our callsign on the aircraft radio.

"Pegasus Bravo, this is La Courtine. Do you read, over?" came the message.

"La Courtine, this is Pegasus Bravo. Go ahead, over," I replied.

"Pegasus Bravo. We have a task for you." I re-checked our fuel state before responding.

"Roger, La Courtine," I said. "We can offer you 40 minutes on task before refuelling."

"Standby, Pegasus Bravo," came the reply. "Grid reference follows. Confirm ready to copy?"

"Affirmative," I said. "Go ahead." Mike wrote down the grid reference, and checked the position on his map.

"It's not far from here," he said.

"Pegasus Bravo from La Courtine. There's trouble with a vehicle at that grid reference. Your assistance appreciated if possible."

"Roger," I replied. "We'll see what we can do."

I changed course, and re-checked my own map as Mike gave directions towards the grid reference. We had about five miles to fly to the position given. The area was still hilly, but unlike some of the more mountainous territory further south, there were cultivated fields in the vicinity.

"Grid reference on the nose, range one mile," called Mike eventually. Ahead, I could see a flat marshy area. A group of military vehicles soon came into view. Two of the vehicles were separated from the rest. A number of uniformed figures stood by one of the vehicles.

"I'll make an approach to a spot near that group of soldiers," I said to Mike.

"Okey, dokey." Mike's jargon signified that he was in a good mood. He gave directions as I came to a hover near the scene of trouble.

"Your position is good," said Mike. "Down ten," Mike continued to give instructions as I lowered the collective lever. Just as the wheels were about to touch the surface, he said: "Maintain your height."

I kept the Wessex in a low hover, whilst Mike checked the area all around. "It's very boggy here," he said after a while. "I suggest you don't land."

"OK," I replied. "I'll hold it in a low hover, while you nip out to speak to those guys."

"Roger," said Mike. He then unplugged his head-set, and I noticed he struggled slightly as he walked through the boggy ground towards the soldiers. Eventually he returned to the helicopter.

Mike attempted to wipe off mud from his boots before entering the helicopter. At length, he climbed aboard, and plugged-in his head-set. Then he said breathlessly: "The prat driving that Land Rover didn't bother to check the terrain properly. As a result, he's bogged down."

"And the four ton truck over there?" I asked.

"The prat driving that one," continued Mike. "Drove over to rescue the prat driving the Land Rover. Now both prats are bogged down."

"Well we can't do anything about the four tonner," I said. "He's too heavy. The Land Rover's also too heavy just to lift up. But if we could connect a tow-rope between the rear of the vehicle and our underslung load hook, we could have a go at towing." Underneath the Wessex was a special hook arrangement. It was designed for load-lifting, but I felt it could be adapted for the present task.

"We could give it a go," said Mike. "They've got a tow-rope."

Slowly, I moved the helicopter closer to the Land Rover. Mike made hand signals to the ground party. The soldiers ran beneath the helicopter to connect the tow-rope. Mike then gave me directions to move behind the vehicle as the slack in the rope was taken up. I eased backwards steadily; any snatching of the rope could have been dangerous. At length, I noticed the effect on the flying controls when the rope reached full stretch.

"Right," said Mike. "The rope is now taught. Continue in that direction very slowly." Gradually I pulled on the collective lever, at the same time I monitored the torque gauge on the instrument panel. Soon the gauge was at the maximum permitted limit. The whole machine felt strained, and flew at a peculiar, unnatural angle as it struggled.

After a while, I said to Mike: "It's not working. Is the Land Rover driver doing anything while we pull?"

"No," replied Mike. "He's just sitting there in a trance." I reduced the torque, and then lowered the Wessex to a low hover while Mike relayed instructions to the driver. After that, I returned to our previous position, and applied torque as we resumed our pull. Suddenly, we felt a slight move backwards.

"That's it!" said Mike. "Something's happening." Slowly, the combined efforts of our pulling, and the vehicle driving, allowed progression towards the firm surface at the edge of the field. Eventually, the helicopter's torque began to reduce as the Land Rover gathered momentum. As we reached the side of the field, Mike said: "Hold it there." I maintained the Wessex in a hover, while Mike continued to monitor proceedings beneath.

"You're clear below," said Mike at length. "Down five... four... three... two... steady Steady. Hold that height." He beckoned in a soldier to unhook the tow-rope.

When the tow-rope had been detached from our hook, I said to Mike; "We'll have to return to La Courtine now. We're getting quite low on fuel."

"Roger," replied Mike. "The soldier below has just left the area of down-wash. You're clear above and behind."

We waved at the soldiers as they indicated grateful thumbs-up signs. Then I turned away from the rescued Land Rover, and flew off towards La Courtine. Eventually, I called on the radio.

"La Courtine this is Pegasus Bravo, do you read? Over." After a short pause, the radio operator at La Courtine called back.

"Pegasus Bravo, this is La Courtine. Go ahead, over."

"Roger," I replied. "One Land Rover has been rescued. We're now returning for refuel."

"Well done, Pegasus Bravo," replied La Courtine. "Refuelling vehicle is ready." The radio operator then paused before adding. "By the way, we've organised a banquet for you this evening."

"Sounds good," I said. "Thank you."

After closing down the Wessex, and supervising the refuel, Mike and I walked over to the Operations set-up.

"What's this about a banquet?" we asked the army co-ordinating Officer.

"Oh, yeah," he replied. "There's something been fixed up in the local village café for tonight. The Mayor's going to be there and everything."

"Excellent," said Mike and I.

"What about the dancing girls?" asked Mike.

"You should be so lucky!" quipped the army Officer.

As the evening approached, and we had changed from our aircrew clothing, an army mini-bus arrived to drive us to the local village café. The crew of the other Wessex were picked up, in addition to Mike and myself. At length, we arrived at the village, and the driver pointed out the café. The four aircrew then left the vehicle, to make their way by foot. We had to cross a village square, before walking down a narrow cobbled street.

Just as we were making our way across the square, a French voice called after us: "*Messieurs! Excusez-mois! Vous êtes Anglais, je crois?*" We all stopped, and turned round.

"*Mais oui!*" I replied. "*Nous sommes Anglais.*"

"*Excusez mois. Excusez mois!*" a short man paced across the square to where we stood. "I'm sorry. I'm sorry!" he had reverted to English. "My English…*c'est mauvais*…" he continued. "But please to come with me for one moment." The four of us looked at each other in some surprise, then we followed the Frenchman. He led us into a small sweet shop, and called out as we entered.

"Please…to meet my *femme*…my *deux filles*…" we shook hands warmly with his wife and two attractive daughters. Then the shop-keeper produced something from the wall at the back of his shop, and held it up for us to inspect.

"It's a battered old watch," remarked Mike.

"*Oui…oui,*" the shop-keeper nodded vigorously. "*Ma montre…* she save my life." He then carefully pointed out the position the watch had been on his chest when it had deflected the German bullet. The four aircrew were intrigued as they examined the genuine artefact. The watch had been compressed almost into two halves by the force of the impact. The precious, yet valueless, item aroused emotion in the family as we perused it. One of the daughters placed her arm around her father. We sensed a poignant atmosphere. The family looked at each other, and at their visitors. We all felt conscious of sadness mixed with compassion.

At length, the shop-keeper said: *"Merci*...thank you. Please to take these. *Pour vos enfants."* He handed to each of us a tin containing local speciality sweets. We shook hands again with the family, and waved back at them as we continued our walk towards the café.

On entering the café, we were quickly taken to a room at the back which had been specially set-up. A long table had been laid, and a number of people were in the room. There was a hubbub of voices, and vigorous hand-shaking took place as we entered. We were eventually invited to take our places at the table. Opposite me sat Mike, who appeared to demur at the sight of the snails produced for our 'starter'.

"Just be grateful it's not hard-tack biscuits," I said to him.

The meal, generous and appetising to most, continued in a good-humoured atmosphere. The language barriers seemed to be eased by the flow of carefully selected wines. Eventually, the Mayor stood up.

"*Mesdames, Messieurs*... Ladies and Gentlemen," said the Mayor. "My English: she is not so good. I therefore make this... how you say... *concis*. Everybody... we are very glad to see you. The War... she is over now. *Mais chose terrible*... terrible things happened in this *domain*... this area. Here, we are not so far from *Vichy*. Things happened which you could... *peut-etre*... maybe... never understand. But we always had hope... because of *General de Gaulle. Toujours*... always... we thank Great Britain for helping him give... true Frenchmen... *espere*... hope during those... dark days. *Mon peuples*... my people want to say... *Merci*... from *le coeur*... the heart. Now we look *a futur*... please to stand. I give *un toast à la Reine Elizabeth!*"

There was a shuffling of chairs as the assembly stood up for the loyal toast. It fell to me to make a brief reply to the Mayor's speech, and to invite raised glasses to toast *Monsieur le Président de la République*.

The evening had been a mixture of enjoyment and emotion. Eventually, the assembled guests shook hands once more, and started to make their way out of the café. As we stepped outside, we became aware of the Autumn chill. The street was dark and forlorn in contrast with the atmosphere inside the café.

"Autumn's in the air," Mike shivered as we walked along the cobbled street back towards the village square. We continued to converse as we strolled, and he suddenly asked me: "You leave the Service soon, don't you?"

"In a few months' time," I replied.

"What do you plan to do?"

"Hopefully, I'll get a job as a civilian pilot," I said.

"Hm," grunted Mike. "It'll be a rather different life to this one, that's for sure." He seemed reflective; after some while, he sighed: "At least you've done your bit for Queen and Country. They need good people when the going gets tough. As the old Mayor was trying to say, just think where we'd be now if their generation had all capitulated to Hitler's Germany."

We walked at a leisurely pace up the cobbled village street. At length, I asked Mike: "Do you remember that survival exercise we did together in Bavaria?"

"Certainly," he replied. "As if I could forget it!"

We fell quiet for a moment, and continued to walk slowly along the rough surface of the street. Eventually, I said : "It's just that at times like that, and also this evening…" I hesitated for a moment. "Well, somehow…it makes you think, doesn't it?"

Mike was still pensive. Finally, however, he said: "It makes me think that – if nothing else – we've had a varied lifestyle. What's that saying about variety being the spice of…?"

We were just at the end of the small cobbled street; we approached the village square. It was quite late in the evening. Most of the local houses had been locked-up for the night. Around us the atmosphere was hushed. I glanced at Mike; he still appeared to be absorbed in thought. Then I broke the silence.

"The spice of flight, I think you mean."

Also published by Woodfield...
The following titles are all available in our unique high-quality softback format

RAF HUMOUR

Bawdy Ballads & Dirty Ditties of the RAF – A huge collection of the bawdy songs and rude recitations beloved by RAF personnel in WW2. Certain to amuse any RAF veteran. Uncensored – so strictly adults only! *"Not for the frail, the fraightfully posh or proper gels – but great fun for everyone else!"* **£9.95**

Upside Down Nothing on the Clock – Dozens of jokes and anecdotes contributed by RAF personnel from AC2s to the top brass... still one of our best sellers. *"Highly enjoyable."* **£6.00**

Upside Down Again! Our second great collection of RAF jokes, funny stories and anecdotes – a great gift for those with a high-flying sense of humour! *"Very funny indeed."* **£6.00**

Was It Like This For You? A feast of humorous reminiscences & cartoons depicting the more comical aspects of life in the RAF. *"Will bring back many happy memories. Highly recommended."* **£6.00**

MILITARY MEMOIRS & HISTORIES – THE POST-WAR PERIOD

The Spice of Flight by **Richard Pike** Former RAF pilot of Lightnings, Phantoms and later helicopters with 56, 43(F) & 19 Sqns delivers a fascinating account of RAF flying in the 60s & 70s. **£9.95**

Meteor Eject! by **Nick Carter** Former 257 Sqn pilot [1950s] recalls the early days of RAF jets and his many adventures flying Meteors, including one very lucky escape via a Mk.1 Irvin ejector seat... **£9.95**

I Have Control... by **Edward Cartner** A former RAF Parachute instructor humorously recalls the many mishaps, blunders and faux-pas of his military career. *Superb writing; very amusing indeed.* **£9.95**

Return to Gan by **Michael Butler** Light-hearted account of life at RAF Gan in 1960 and the founding of 'Radio Gan'. *Will delight those who also served at this remote RAF outpost in the Indian Ocean.* **£12.00**

KOREA: We Lived They Died by **Alan Carter** Former soldier with Duke of Wellington's Regt reveals the appalling truth of front-line life for British troops in this now forgotten war. *Very funny in places too.* **£9.95**

MILITARY MEMOIRS & HISTORIES – WORLD WAR 1 & 2

Once a Cameron Highlander by **Robert Burns** The autobiography of Robert Burns, a veteran of the Battle of the Somme who lived to be 104 and thus became the oldest surviving Cameron Highlander. A fascinating account of his WW1 experiences, his later eventful life in show business and his celebrity status as a centenarian when he was the guest of honour at many glittering social occasions. **£9.95**

2297: A POWs Story by **John Lawrence** Taken prisoner at Dunkirk, the author endured 5 years as a POW at Lamsdorf, Jagendorf, Posen and elsewhere. Very interesting & delightfully Illustrated. £6.00

A Bird Over Berlin by **Tony Bird DFC** Lancaster pilot with 61 Sqn tells a remarkable tale of survival against all the odds during raids on the German capital. *"An almost miraculous sequence of events."* **£9.95**

A Shillingsworth of Promises by **Fred Hitchcock.** Delightfully funny and ribald memoirs of an erk during the wartime years and beyond in UK and Egypt. A very entertaining read. **£9.95**

A Journey from Blandford by **B.A. Jones.** The wartime exploits of this motorcycle dispatch rider/MT driver began at Blandford Camp but involved Dunkirk, the Middle East, D-Day and beyond... **£9.95**

A Lighter Shade of Pale Blue by **Reg O'Neil** A former Radar Operator recalls his WW2 service in Malta and Italy with 16004 AMES a front-line mobile radar unit. *'Interesting, Informative and amusing.'* **£9.95**

Beaufighters BOAC & Me by **Sam Wright** – WW2 Beaufighter navigator served full tour with 254 Sqn and was later seconded to BOAC on overseas routes. *'Captures the spirit of the mighty Beaufighter'* **£9.95**

Coastal Command Pilot by **Ted Rayner.** Former Hudson pilot's outstanding account of WW2 Coastal Command operations from Thornaby, St Eval, Wick, Iceland and Greenland. **£8.00**

CYRIL WILD: The Tall Man Who Never Slept by **James Bradley**. Biography of a remarkable Japanese-speaking British Army officer who helped many POWs survive on the infamous Burma railway. **£9.95**

Desert War Diary by **John Walton** Diary and photos recording the activities of the Hurricanes and personnel of 213 Squadron during WW2 in Cyprus and Egypt. *"Informative and entertaining."* **£9.95**

Fiji to Balkan Skies by **Dennis McCaig** Spitfire/Mustang pilot recalls eventful WW2 operations over the Adriatic/Balkans with 249 Sqn in 43/44. *'A rip-roaring real-life adventure, splendidly written.'* **£9.95**

From Horses to Chieftains by **Richard Napier** An Army career from Egypt in 1935 with 8th Hussars through WW2 with 7th Armoured division (Desert Rats) and on to the 1960s is entertainingly remembered. **£9.95**

Get Some In! by **Mervyn Base** The life & times of a WW2 RAF Bomb Disposal expert £9.95

Harry: An Evacuee's Story by **Harry Collins** The sometimes harrowing but eventually heart-warming story of the trials and tribulations of one young evacuee sent from Stockport to Canada in WW2. £9.95

Operation Pharos by **Ken Rosam** The story of the Cocos Keeling islands and of operations from the RAF's secret bomber base/staging post there during WW2. *'A fascinating slice of RAF history.'* £9.95

Over Hell & High Water by **Les Parsons** This WW2 navigator survived 31 ops on Lancasters with 622 Sqn, then flew Liberators in Far East with 99 Sqn. An exceptional tale of 'double jeopardy'! £9.95

Ploughs, Planes & Palliasses by **Percy Carruthers** Entertaining recollections of an RAF pilot who flew Baltimores in Egypt with 223 Squadron and was later a POW at Stalag Luft 1 & 6. £9.95

Training for Triumph by **Tom Docherty** This encyclopaedic work contains details about every training facility operated by the RAF during World War 2. *'An impressive achievement.'* £12.00

Memoirs of a 'Goldfish' by **Jim Burtt-Smith** This former Wellington pilot is a founder member and president of the Goldfish Club, exclusively for aviators who have force-landed into water. Jim describes 'ditching' in the North Sea along with many more entertaining episodes from his eventful War and later life. £9.95

Pathfinder Force Balkans by **Geoff Curtis** Pathfinder F/Engineer saw action over Germany & Italy before baling out over Hungary. POW in Komarno, Stalags 17a & 17b. Amazing catalogue of adventures. £9.95

Per Ardua Pro Patria by **Dennis Wiltshire** Humour and tragedy are interwoven in these unassuming autobiographical observations of a former Lancaster Flight Engineer who later worked for NASA. £9.95

Pacifist to Glider Pilot by **Alec Waldron** This son of Plymouth Brethren parents renounced pacifism and went on to pilot Hadrian & Horsa gliders at Sicily & Arnhem. *'A remarkable story, well told.'* £9.95

The RAF & Me by **Gordon Frost** Stirling navigator recalls ops with 570 Sqn from RAF Harwell, including 'Market-Garden' 'Varsity' and others. *'A salute to the mighty Stirling and its valiant crews.'* £9.95

Railway to Runway by **Leslie Harris** Wartime diary & letters of a Halifax Observer – later killed in action with 76 Sqn in 1943 – capture the spirit of the wartime RAF as recorded by a 19 year old airman. £9.95

A Lighter Shade of Pale Blue by **Reg O'Neil** Tales of the eventful war of a Type16 Mobile Radar Op in UK, Malta, Corsica & Italy with 16004 AMES. *'Interesting and often very funny.'* £9.95

Just a Survivor by **Phil Potts** Former Lancaster navigator with 103 Sqn tells his remarkable tale of survival against the odds in the air and as a POW. *'An enlightening and highly agreeable account.'* £9.95

RAF/UXB The Story of RAF Bomb Disposal Stories contributed by wartime RAF BD veterans that will surprise and educate the uninitiated. *"Amazing stories of very brave men."* £9.95

UXB Vol 2 More unusual and gripping tales of bomb disposal in WW2 and after. £9.95

Nobody's Hero by **Bernard Hart-Hallam**. An RAF SP's extraordinary adventures with 2TAF on D-Day and beyond in France, Belgium & Germany. *"Unique and frequently surprising wartime tales."* £9.95

No Brylcreem, No Medals by **Jack Hambleton** – RAF MT driver's splendid account of his wartime escapades in England, Shetlands & Middle East blends comic/tragic aspectsof war in uniquely entertaining way. £8.00

Un Grand Bordel by **Norman Lee & Geoffrey French** – WW2 tail gunner's fascinating account of his adventures with the French Secret Army after being shot down is both funny and highly eventful. £9.95

While Others Slept by **Eric Woods** The story of Bomber Command's early years WW2 by a Hampden navigator who completed a tour with 144 Squadron. *'Full of valuable historical detail.'* £9.95

WOMEN in WORLD WAR TWO

Radar Days by **Gwen Arnold** Delightful evocation of life in the wartime WAAF by a former Radar Operator at Bawdsey Manor RDF Station, Suffolk. *"Amusing, charming and affectionate."* £9.95

Corduroy Days by **Josephine Duggan-Rees** Ex-Land Girl's warm-hearted and amusing recollections of wartime years spent on farms in the New Forest area. *"Funny, nostalgic and very well written."* £9.95

Lambs in Blue by **Rebecca Barnett**. Revealing account of the wartime lives and loves of a group of WAAFs posted to the tropical paradise of Ceylon. *"A highly congenial WW2 chronicle."* £9.95

A WAAF at War by **Diana Lindo** Former MT driver's charming evocation of life in the WAAF will bring back happy memories to all those who also served in World War 2. *"Nostalgic and good-natured."* £9.95

'Ernie' by **Celia Savage** A daughter's quest to discover the truth about the death of her RAF father, a Halifax navigator with 149 Sqn who died in WW2 when she was just 6 years old. £9.95

Searching in the Dark by **Peggy Butler** The amusing diary of a former WAAF radar operator 1942-1946 written when she was 19 yrs old and serving at Bawdsey Manor RDF station in Suffolk £9.95

MEMOIRS & HISTORIES – NURSING AND HEALTHCARE

Call and Ambulance by **Alan Crosskill** – A former ambulance driver recalls a number of episodes from his eventful career in the 1960s/70s. The stories are all extremely amusing and entertaining, so much so that he has already been hailed as the 'James Herriot of Ambulancemen'... **£9.95**

Occupation Nurse by **Peter & Mary Birchenall**. Peter is a Lecturer in Health Studies and Mary, a Senior Lecturer in Nursing at a Northern University. Their book describes the unique achievement of a group of untrained young women who, under the supervision of three trained Matrons, provided health care for the sick at the island of Guernsey's only hospital during the German occupation of 1940-45. **£9.95**

Just Visiting... by **Molly Corbally**. An absolutely charming book by a former Health Visitor, who brilliantly depicts the many colourful characters she worked with and recalls many entertaining episodes from her long career – the 1940s to the 1970s – in the rural villages near Kenilworth in the West Midlands. **£9.95**

FICTION

Trace of Calcium by **David Barnett** – When a mild-mannered commuter comes to the aid of a young woman in trouble, he becomes implicated in a murder and must use all his resources to clear his name. **£9.95**

Double Time by **David Barnett** – A light-hearted time-travel fantasy in which a bookmaker tries to use a time machine to make his fortune with hilarious consequences. **£9.95**

Last Sunset by **AA Painter** A nautical thriller set in the world of international yachting. A middle aged yachtsman becomes accidentally embroiled with smugglers, pirates and a very sexy young lady... **£9.95**

Retribution by **Mike Jupp** A brilliantly illustrated and very funny comedy/fantasy novel for adults and older children featuring bizarre goings-on in a quiet seaside town. A glorious mixture of knockabout comedy and escapist fantasy with an environmental message thrown in for good measure. An absolute delight from start to finish. *"A Brilliant tribute to his home town."* Mike Read, BBC **£9.95**

MISCELLANEOUS SUBJECTS

Just a Butcher's Boy by **Christopher Bolton** Charming account of small town life in the 1950s in the rural Leiston, Suffolk and idyllic summers spent with grandparents who owned the local butcher's shop. **£5.95**

Impress of Eternity by **Paul McNamee** A personal investigation into the authenticity of the Turin Shroud. A former shcoolmaster examines the evidence and comes to a startling conclusion. **£5.95**

Making a Successful Best Man's Speech An indispensable aid to anyone who feels nervous about making a wedding speech. Tells you what to say and how to remember it. **£5.95**

Near & Yet So Far by **Audrey Truswell** The founder of an animal rescue charity tells charming and heart-warming tales of the rescue and rehabilitation of many four-legged friends in need. **£9.95**

Reputedly Haunted Inns of the Chilterns & Thames Valley by **Roger Long** – A light hearted look at pubs & the paranormal in the Heart of England **£5.95**

A Selection of London's Most Interesting Pubs by **David Gammell** – A personal selection of London's most unusual and historic hostelries with instructions how to find them. **£5.95**

Unknown to History and Fame by **Brenda Dixon** – Charming portrait of Victorian life in the West Sussex village of Walberton via the writings of Charles Ayling, a resident of the village, whose reports on local events were a popular feature in *The West Sussex Gazette* over many years during the Victorian era. **£9.95**

Woodfield books are available direct from the publishers by mail order as well as via all usual retail channels...

Telephone your orders to (**+44** if outside UK) **01243** 821234
Fax orders your to (**+44** if outside UK) **01243** 821757
All major credit cards accepted.

Visit our website for full details of all our titles – instant online ordering is also available at **www.woodfieldpublishing.com**

Woodfield Publishing

BABSHAM LANE ~ BOGNOR REGIS
WEST SUSSEX ~ ENGLAND
PO21 5EL

www.woodfieldpublishing.com